An Introduction to Microscopy by Means of Light, Electrons, X-Rays, or Ultrasound

Field-ion micrograph of a platinum crystal by the late Professor
Erwin W. Müller of The Pennsylvania State University, by courtesy
of Professor T. T. Tsong, Department of Physics, The Pennsylvania
State University.

An Introduction to Microscopy by Means of Light, Electrons, X-Rays, or Ultrasound

Theodore George Rochow

North Carolina State University at Raleigh
Raleigh, North Carolina

and

Eugene George Rochow

Harvard University
Cambridge, Massachusetts

PLENUM PRESS · NEW YORK AND LONDON

Library of Congress Cataloging in Publication Data

Rochow, T   G
  An introduction to microscopy by means of light, electrons, x-rays, or ultrasound.

  Includes bibliographical references and index.
  1. Microscope and microscopy. I. Rochow, Eugene George, 1909-   joint author. II.
Title. [DNLM: 1. Microscopy. 2. Microscopy, Electron. 3. Radiation, Ionizing. 4. Ultrasonics.
QH205.2 R681i]
QH205.2.R63   502'.8   78-7529
ISBN 0-306-31111-9

©1978 Plenum Press, New York
A Division of Plenum Publishing Corporation
227 West 17th Street, New York, N.Y. 10011

Printed in the United States of America

In honor of
the succession of teachers of microscopy at Cornell University

SIMON HENRY GAGE
EMILE MONNIN CHAMOT
CLYDE WALTER MASON
and
GEORGE GOSSON COCKS

Preface

Many people look upon a microscope as a mere instrument[1]; to them microscopy is instrumentation. Other people consider a microscope to be simply an aid to the eye; to them microscopy is primarily an expansion of macroscopy. In actuality, microscopy is both objective and subjective; it is seeing through an instrument by means of the eye, and more importantly, the brain. The function of the brain is to interpret the eye's image in terms of the object's structure. Thought and experience are required to distinguish structure from artifact. It is said that Galileo (1564–1642) had his associates first look through his telescope–microscope at very familiar objects to convince them that the image was a true representation of the object. Then he would have them proceed to hitherto unknown worlds too far or too small to be seen with the unaided eye.

Since Galileo's time, light microscopes have been improved so much that performance is now very close to theoretical limits. Electron microscopes have been developed in the last four decades to exhibit thousands of times the resolving power of the light microscope. Through the news media everyone is made aware of the marvelous microscopical accomplishments in imagery. However, little or no hint is given as to what parts of the image are derived from the specimen itself and what parts are from the instrumentation, to say nothing of the changes made during preparation of the specimen. It is the purpose of this book to point out the limitations as well as the advantages of various microscopes and methods to enable the reader to make his own interpretations and draw his own conclusions.

Many writers consider magnification as a primary function of any microscope. To the present authors useful magnification is merely incidental to the *resolution* of detail in the object as imaged on the retina or in the photomicrograph. This approach puts a somewhat different emphasis on the various attributes contributing to microscopical visibility.

The purpose of microscopy is to serve every scientific technological discipline on earth. This book honors the biologist Gage,[2] the chemist Chamot,[2] the engineer Mason,[3] and the electron microscopist Cocks[4] because they deliberately crossed interdisciplinary lines in their teaching of old and new principles. Learning, they knew, can proceed from any area of science to any other. For example, Pasteur[5] started his illustrious microscopical career in chemistry, proceeded to biochemistry, and concluded with medicine. A contemporary, Henry Clifton Sorby,[6] used the microscope to study rocks, ores, meteorites, iron and steel, fossils, and living creatures—crossing the nebulous lines of science continually.

The authors are greatly indebted to North Carolina State University, the American Cyanamid Company, the American Society for Testing and Materials, Cornell University, Harvard University, and The Pennsylvania State University. We wish to express our gratitude to the many colleagues, students, co-workers, and peers who have helped us to develop the text.

The authors thank Dr. Paul A. Tucker, Jr., of North Carolina State University at Raleigh, for many contributions and comments regarding light, electron, and scanning acoustic microscopy. Acknowledgment is also gratefully made to the following people in Instructional Technology Services (ITS), School of Textiles of the same University: Thomas L. Russell, in charge; Mark L. Bowen, graphic designer who produced most of the line drawings; and Reed A. Petty, who photographed Figure 5.8 in color.

We especially thank Elizabeth Cook Rochow and Helen Smith Rochow for their part in editing and typing the manuscript and index.

Comments and criticisms have been so helpful in the past that we welcome further suggestions from our readers.

Raleigh, North Carolina   Theodore G. Rochow
                                                              Eugene G. Rochow

Contents

3. Simple and Compound Microscopes

4. Compound Microscopes Using Reflected Light

5. Microscopy with Polarized Light

6. The Microscopical Properties of Fibers

7. Microscopical Properties of Crystals

8. Photomicrography

9. Contrast: Phase, Amplitude, and Color

10. Interferometry in Microscopy

11. Microscopical Stages

12.   Transmission Electron Microscopy

13.   Scanning Electron Microscopy

14. Field-Emission Microscopes

15. X-Ray Microscopy

16. Acoustic Microscopy

A Brief History of Microscopy

1.1. INTRODUCTION

All kinds of microscopy have a common beginning in mankind's intellectual goal to see better. Visible light was the first medium and visibility was limited to that of the unaided eye. During the first century A.D. it was discovered[1] that by looking through a clear spherical flask filled with clear water "letters however small and dim are comparatively large and distinct."[2] These are the words of Seneca as an old man in about 60 A.D. During the next dozen centuries spherical segments of clear minerals were set in frames as eyeglasses to help older people see better. In about the year 1300 clear silicate glass was being made in Italy, and superstition against its use in eyeglasses for far-sightedness was overcome.[3–5] By the sixteenth century concave lenses were also for sale to help near-sighted people. The availability of both convex and concave lenses led the Dutch to the empirical combination of the two as a crude compound microscope. Just who invented the Dutch microscope is still not known. Contemporary witnesses attributed the construction around 1590 to Zacharias Janssen,[6] but a modern study (1967)[7] reveals that he was born in 1588! Possibly his father, Hans Janssen (who died in 1593), or John Lippershey, or James Metius[6] first evolved the combination of two lenses. The principle of a concave lens serving as an amplifier to the objective is still in use today to give a flat field in some instruments for projection and photomicrography.

In 1611 the Dutch microscope came to the attention of the German mathematician and astronomer Johann Kepler (1571–1630). He was quick to visualize that enlargement could also be obtained by combining a convex (instead of concave) ocular with the convex objective, but that the image would appear inverted rather than erect. The Kepler ocular serves to enlarge the real image of the objective, which is the principle of the modern light microscope.[2] Kepler also showed the dioptrics of the eye, and explained how convex spectacle lenses correct for hyperopia and how concave lenses correct for myopia. By 1619 Cornelius Drebbel in London had made a compound microscope with a biconvex eye lens and a plane–convex objective the size of a small cherry, stopped down by a diaphragm the size of a thick needle. It gave an inverted image.[7] In May 1624, Galileo received the Drebbel microscope, demonstrated how it worked, and indicated that he had made similar, but better, "occhiali" (eyeglasses). In the same year Galileo ground lenses, mounted them into microscopes, and sent them to friends.[7]

The word *microscope* was coined by Giovanni Faber on April 13, 1625.[6] It meant the same then as it does today—an optical instrument capable of producing a magnified image of a small object[8]—but today, radiations other than light may be used for illumination.

Robert Hooke (1635–1703), as Curator of Experiments for the Royal Society of England, possessed a compound microscope built by Christopher Cock. Hooke experimented a great deal with his microscope,[9,10] taking out and putting back the "middle glass" (field lens) and providing himself with a condensing lens (a globe filled with clear brine). Because of his devotion to microscopical observations for the Royal Society, King Charles II requested a "handsome book" of Hooke's reports. Hooke's skill and early training as an artist were manifested in the beautiful illustrations of chiefly biological materials in the commissioned book, *Micrographia*. Originally published in 1665, Hooke's classic book has been reprinted, as evidenced by Figure 1.1.[11]

In 1646 Althanasius Kircher (1601–1680) described and used a simple microscope "about the size of a thumb," consisting of a single convex lens separated from a plane glass used for holding the specimen,[1] in the manner of the simple microscope–micrometer of today.

Anton van Leeuwenhoek (1632–1723) made and used microscopes consisting of very small, high-powered, single lenses, perfectly ground into strong convex surfaces.[9,10] Starting about 1670, he made hundreds of microscopes, only a few of which have been preserved. Of these, the one at the University of Utrecht is the best: Although scratched, the lens still resolves 1.4 μm at 270×.[7] During his long career, Leeuwenhoek's curiosity led him to look everywhere in the microcosm. He observed human capillaries and red blood cells with increasing care and detail; he discovered the one-celled animals now called protozoa, and in 1683 he described what we now believe were bacteria.[4] Leeuwenhoek did not go beyond observing and describing, but in these he excelled. His secret of grinding and polishing excellent lenses smaller than a pin's head died with him.

Throughout the seventeenth century, experiments in empirically

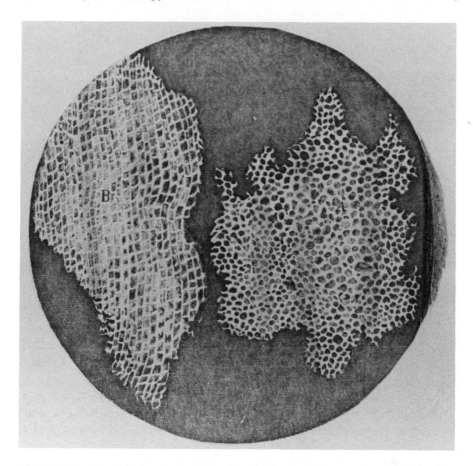

FIGURE 1.1. Hooke's drawings of cork showing the "cells," a word he used because the tiny spaces reminded him of monks' cells.[11]

arranging two, three, or more lenses in a tube were continued, but the salient advance for us today was the advent of the Huygens eyepiece, invented by the Dutch physicist and astronomer Christiaan Huygens (1629–1695) and his brother. The Huygens eyepiece has two planoconvex lenses, each with the convex side facing the objective. The lower (field) lens modifies the real image from the objective so as to give a brighter but smaller image.[12] The Huygens design is still the principal one for microscopical eyepieces of magnifications of 10× and under. Reticles of linear, areal, or angular scales are used now in such eyepieces for micrometry, as they were in Huygens' day.

The hand-held crude wooden or iron microscopes of the seventeenth century evolved into the elegant eighteenth-century table microscopes,[13] made chiefly of brass. The importance of the evolution of microscopical stands centers on the illumination of the object. Mirrors were developed to transmit light through transparent objects, while others were made to reflect light from opaque objects.

1.2. CORRECTIONS FOR ABERRATIONS

During the seventeenth century there was little optical improvement because the causes of aberrations and artifacts[8] were not understood. However, at the turn of the eighteenth century David Gregory (1661–1708), a friend of Sir Isaac Newton (1642–1727), noted something which had escaped the master: Different kinds of glass spread out the colors of the spectrum to different extents.[4] He suggested that a proper combination of two kinds of glass might produce no spectrum at all. This feat was accomplished commercially by John Dolland, an English optician (1706–1761), and his son Peter. Their achromatic[8] objectives demonstrated the superiority of refracting telescopes over the existing reflecting telescopes, and continued to do so until the twentieth century, when reflectors took over.[4]

Achromatic microscopical objectives are much more difficult to construct than telescopic ones because they are much smaller. The first commercial achromatic objectives appeared in 1824 and were of French manufacture. The early ones were of smaller numerical aperture than the corresponding uncorrected objectives,[2] so they produced less brilliant images and were at first unpopular. However, achromatic objectives of higher aperture were perfected rapidly during the rest of the nineteenth century.[14–20] A vivid example of this progress is provided by comparing Figures 1.2 and 1.3.

Meanwhile, there was corresponding improvement in oculars. In 1782 Jesse Ramsden (1735–1800), an English telescope maker, produced an ocular of new design. Like Huygens' eyepiece, Ramsden's contained the two planoconvex lenses, but the convex sides were turned toward each other and were close together for better correction of spherical aberrations. Both lenses were mounted so that they were above the focal plane of the objective's real image, where the crosshair or micrometer scale normally is placed, so the ocular could magnify both. The Ramsden type of eyepiece is used today for most work requiring an eyepiece magnification of about 12× or greater.[12]

By the beginning of the nineteenth century, microscopists were setting up standards for measuring resolving power,[8] employing natural structures such as the striated scales of butterflies, followed by diatoms and then by artificial rulings.[7,20] Thus it was discovered that

FIGURE 1.2. Gnat's wing as seen with an uncorrected Cuff objective, ca. 1750. Photomicrograph courtesy of Dr. Maria Rooseboom, Director of the National Museum of the History of Science, Leiden, the Netherlands.

resolution increased with angular aperature in the objective and condenser.

One of the chief investigators, Giovanni Battista Amici, combined the qualities of theorist, practical optician, and microscopist. Among his important innovations was the use of a single hemispherical front lens in the microscopical objective (in 1844), a principle that still prevails. In 1830 Joseph J. Lister (1786–1869), father of the famous surgeon, discovered the two aplanatic foci of achromatic doublets. He worked on the problem of the increasing influence of cover-glass thickness when using objectives of increasing aperture. Different solutions to the problem were found. In 1837 Andrew Ross made objectives with movable separation of lenses to correct for the varying thicknesses of cover glasses. A decade later Amici showed that immersing the objective in a liquid of appropriate refractive index could eliminate problems of the cover glass.[7]

Until the early nineteenth century it was thought that perfection of lenses alone would lead to the resolution of infinitely small structures. Then in 1834, the astronomer George B. Airy (1801–1892) showed that

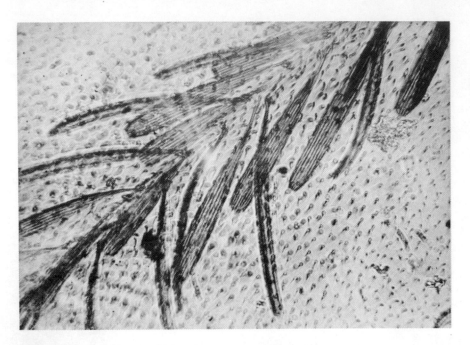

FIGURE 1.3.   Gnat's wing as seen with an achromatic Oberhauser objective, ca. 1850. Photomicrograph courtesy of Dr. Maria Rooseboom, Director of the National Museum for the History of Science, Leiden, the Netherlands.

light from a star would never be focused at a single point, but only in a disk. Therefore, in microscopy there must also be a limiting angular distance between self-luminous objects in order to make them appear separate. This angle would depend on the radius of the lens and the wavelength of light. In 1873 Ernest Abbe (1840–1905) came to essentially the same conclusion for periodic structures, thereby giving practical meaning to the concept of numerical aperture.[8] Abbe worked with the Carl Zeiss Company in Jena in 1846; his many inventions include apochromatic objectives (1885) and their compensating eyepieces. Along with Otto Schott, he succeeded in greatly improving the quality of optical glass.[21] In 1857 the British Royal Microscopical Society introduced a standard size and pitch of screw thread for objectives, but met with difficulty in perfecting taps to be supplied to manufacturers. In an effort to perfect the standard, the American Microscopical Society worked with the Royal Microscopical Society and eventually arrived at a joint standard. Today, American machining practice also is standardized.[22] Virtually every manufacturer in the world now produces objectives and microscopes fitted with "Society" threads, and to this extent interchangeability is ensured.

Standards for eyepieces and substages have not been as universally accepted. However, there is a set of standards for these and other parts under the supervision of the British Standards Institution.[23] In America, interchangeability of microscopical parts is within the scope of the National Committee E-25 on Microscopy in the American Society for Testing and Materials.[24]

1.3. DARK-FIELD MICROSCOPY

It is doubtful that Leeuwenhoek's secret of unique success lay entirely in his craftsmanship in grinding tiny lenses. The biologist, Clifford Dobell (1886–1949) postulated that Leeuwenhoek's secret was in obtaining a dark-field effect: objects seen brightly due to light from the side against a dark background.[1] Two centuries after Leeuwenhoek, J. J. Lister (in 1830) used dark-field microscopy to take advantage of light scattered by surfaces and other discontinuities.[8] Special dark-field condensers for higher-powered objectives were made available by Francis H. Wenham (1823–1908).[2] He made use of a paraboloid condenser and central stop to give a hollow cone of light directed outside the angular cone of the objective.

1.4. POLARIZING MICROSCOPE

The double refraction of calcite was described by Erasmus Bartholin (1626–1697) in 1669. Etienne L. Malus (1725–1812) coined the term "polarized light." In 1828 William Nicol (1768–1851) used calcite to make his famous polarizing and analyzing prisms. Other designs followed, but all plane-polarizing calcite prisms became known as nicols.[8] As a pair (polarizer + analyzer), they are used in a polarimeter to study optical activity[12] of liquids, or in a polariscope to study optical properties of solids.[25] Louis Pasteur used both, together with his microscope, on tartaric acids and their salts to discover a whole new class of chemical isomers.[21,26]

The Englishman Henry Clifton Sorby (1826–1908) employed polarized light microscopically to study thinned sections of limestones and other transparent rocks to identify their mineral constituents.[27] Sorby and Pasteur represent a group of scientists of the golden nineteenth century whose microscopical research went far beyond detailed description of specimens. They postulated mechanisms, drew broad scientific theories, and pointed the way to a variety of technologies.

By the twentieth century the light microscope had reached the limit of highly developed theory for resolving power and contrast not only for visible light but also for the ultraviolet and infrared regions. By the 1930s light microscopy was playing a part in every science and technology. Great professors such as Abbe in Germany had worked with outstanding producers of optical glass, namely, Schott, and of microscopes, namely, Zeiss. Abbe's experiments with phase differences and amplitude contrasts in microscopical images were followed by Zernike's phase-amplitude microscope, for which he won the Nobel prize in 1953.[28]

The development, production, and use of interference microscopes followed a similar pattern. From Newton's general observations of interference phenomena in the seventeenth century, Young developed the fundamental theory of interference in 1801. During the first quarter of the twentieth century, Michelson (1852–1931) developed an interferometer for reflecting objects, and Mach and Zender developed an interferometer for examining transparent objects. By the middle of the century several companies offered interference microscopes.[29,30]

1.5. THE SHORT HISTORY OF ELECTRON MICROSCOPES

Two events in the 1920s brought about the development of the electron microscope. One was the realization from the de Broglie theory (1924) that particles have wave properties and that very short wavelengths (e.g., 0.05 Å) are associated with an electron beam of high energy. The other event was the demonstration by Busch in 1926–1927 that a suitably shaped magnetic field could be used as a lens in electron microscopes.[31,32] Busch and E. Ruska initiated studies of electromagnetic lenses in 1928–1929, and published a description of an electron microscope in 1932. In 1934 Ruska described the construction of his type of electron microscope, which surpassed for the first time the resolution of light microscopes. In 1938 Ruska and von Borries designed and built a practical microscope for the Siemens and Halske Company, but World War II prevented its sale and use outside of Germany.

Independently, at the University of Toronto in Canada, under the supervision of E. F. Burton,[33] A. Prebus and J. Hillier built an electromagnetic electron microscope which they described in 1939. A similar instrument built in the United States was described by C. E. Hall. In 1934 Ladislaus L. Marton built an electron microscope in Brussels with which he took the first electron micrographs of biological objects, such as bacteria. He, with Hillier and Vance at the Radio Corporation of America, helped to build an electron microscope (1940) under the direction of Vladimir K. Zworykin, the inventor of the television picture tube. RCA's first commercial model of this electron microscope went to the American Cyanamid Research Laboratories in December

1940.[34,35,36] In the same year the Columbian Carbon Company built an electron microscope at the University of Toronto under the terms of a Research Fellowship for William A. Ladd, another of Professor Burton's students. The microscope was moved to the Columbian Carbon Research Laboratories in 1941, where pioneering industrial research was performed.[37,38]

In England, Metropolitan Vickers produced a prototype electron microscope (EMI) in 1939, which has been developed into a series of improved models.[39] The Phillips commercial electron microscope was being developed in the Netherlands at the same time. In the 1940s, companies in France, Germany, Japan, and other countries developed and produced this type of microscope, now known as the transmission electron microscope (TEM).

The word "transmission" suggests one of the most important practical problems: obtaining specimens thin enough to allow sufficient transmission of electrons that affect the photographic material satisfactorily without affecting the specimen detrimentally (by heat of absorption of electrons).[39] In the vast science of biology, the development of practical electron microscopy depended primarily upon corresponding improvement of microtomy, so that sections of tissue could be sliced much thinner than those required in light microscopy. Also, differential stains had to be developed on the basis of differential electron absorption by elements of relatively high atomic number, rather than differential light absorption or color. Films, whether in the form of specimens, substrates, or replicas, need to be sufficiently thin. In the 1940s, in both biological and nonbiological sciences, there were also problems of obtaining contrast. "Shadows" were made by preferentially evaporating a metal onto the specimen in a high vacuum. Such problems were complicated by obtaining and maintaining high vacuum in the electron microscope itself. There were other important mutual problems such as maintenance, repair, resolution, magnification,[40] and interpretation.[41] Whereas light microscopists had struggled for centuries over interpretation of images, including macroscopic ones, this problem was newly introduced to electron microscopists in 1945 (Figure 1.4).[41]

As powerful and useful as the transmission electron microscope (TEM) has become since 1940, it is virtually without correction for spherical aberrations. That is, progress toward its potential resolving power is at about the same stage as that of the light microscope in the seventeenth and eighteenth centuries. But Ernst Ruska,[42] Albert Septier,[43] and other electron lens designers are struggling with this problem, as well as that of achieving electrical and mechanical stability. There is now a bright outlook for achieving the theoretical limits of visibility in a much shorter time than it took for the light microscope.

In 1938 M. von Ardenne added scan coils to the TEM. Applications

FIGURE 1.4. Two different orientations (180° apart) of the same photo*macro*graph of depressions in wet sand: (left) apparent depressions, (right) apparent elevations,[41] taken to demonstrate the importance of orientation in the interpretation of images.

are limited to specimens thin enough to transmit electrons so as to activate the photographic medium.[44] Historically, the chief importance of these scan coils may lie in the experience which led to the scanning (reflection) electron microscope, SEM.

1.6. BRIEF HISTORY OF THE SCANNING ELECTRON MICROSCOPE (SEM)

Instead of transmitting the primary electrons to form an image, in 1942 V. K. Zworykin, J. Hillier, and R. L. Snyder of the Radio Corporation of America utilized the *secondary* electrons reflected from the surface of a specimen to produce an image of the topography. The limiting resolution, however, was only 1 μm, less than that of the light microscope (0.1 μm). By reducing the size of the scanning spot and making other improvements, they increased the resolution to 0.05 μm. Further development of the SEM was suspended during World War II.

In 1948 C. W. Oatley at the University of Cambridge became interested in the SEM. He and D. McMullen built one with a resolving power of 0.05 μm. K. C. A. Smith (1956) made several technological improvements, and T. E. Everhart and R. F. M. Thornley (1960) made use of a light pipe to reduce noise. R. F. W. Pease (1963) and W. C. Nixon (1965) produced a prototype of the Cambridge Scientific Instruments' Mark I.[44,45]

Zworykin, Hillier, and Snyder in 1942 tried a cold field-emission sharp cathode as the source of electrons in their experimental SEM in order to reduce the size and improve the intensity of the source. Instability, however, forced these experimenters to return to the thermionic electron gun. The cold-emission point source was finally improved by A. V. Crewe in 1969 to the point of successful application in the SEM. Another type of electron gun, developed by A. N. Broers, employs a heated, pointed rod of lanthanum hexaboride (LaB_6) because it is brighter and lasts longer than a tungsten filament. The requirement of a higher vacuum, however, limits incorporation of the LaB_6 gun in some types of SEM.[31]

Advances in the sixties and seventies have involved contrast mechanisms not available in other types of instruments. Better crystallographic contrast was produced by crystal orientation, lattice orientation, and lattice interactions with primary beams by D. G. Coates in 1967.[44]

In a reflection microscope the contrast between features is often too low, but it may be enhanced by processing the signal. Early processing was done by nonlinear amplification or differential amplification. Derivative signal processing (differentiation) was introduced in 1970 and 1974. Image storage circuits have been developed so that one can observe the image and/or operate on it off-line. Grain sizes, physical-chemical phases, and other analytical features are emphasized by computer evaluation and scanning electron microscopical images (CESEMI). In fact, the computer has interacted with the SEM. The great depth of focus of the SEM allows for quantitative evaluation of topography of specimens. Direct stereo viewing has been described elsewhere.[44]

1.7. THE ELECTRON-PROBE MICROANALYZER (EPMA)

A close relative of the scanning-electron microscope is the scanning-electron-probe microanalyzer. Instead of recording the scattered electrons, the microanalyzer records the *emitted x-rays* and sorts them according to wavelength by use of a Bragg spectrophotometer. A quantitative analysis is made of a chemical element as the scanner picks up the distribution of the element. At the same time, an enlarged image may be displayed if the x-ray microspectrometer is part of a scanning electron microscopical system.[44]

In 1913 H. Moseley found that the frequency of emitted characteristic x-ray radiation is a function of the atomic number of the emitting element. This led to x-ray spectrochemical analysis in terms of the chemical elements present. The electron microanalyzer of a tiny area ($\approx 1\ \mu m^2$) located by means of a light microscope was patented by L. L. Marton (1941) and J. Hillier (1947). In 1949 R. Castaing, under the direction of A. Guinier, built and described an electron microprobe. In 1951 Castaing developed the fundamental calculations for quantitative x-ray analysis by means of the electron microprobe.[44]

In 1956 V. E. Cosslett and P. Duncumb built the first scanning-electron microprobe, sweeping the beam of electrons across the specimen's surface in a raster, as is done in the SEM. They used the backscattered electron signal to modulate the brightness of a cathode-ray tube, sweeping in synchronism with the electron probe. They also used the x-ray signal to modulate the brightness so as to obtain a scanned image showing the lateral distribution of a particular element.[44]

An especially important development has been the use of diffracting crystals which have large interplanar spacings in measuring long wavelength x-rays from light elements: F, O, N, C, and B.[44]

Another development has been the detection and measurement of any color of visible light (cathodoluminescence) produced by the electron probe. Such luminescence is associated with impurities in minerals or other crystals. Likewise, photon radiation produced by the recombination of excess hole–electron pairs in a semiconductor can be studied.[44]

Increased use of the computer in conjunction with the EPMA has greatly improved the quality and quantity of the data obtained. Computers have been programmed to convert x-ray intensity ratios into chemical compositions, and are now being used to automate repetitive EPMA analyses.[44]

Recently, solid-state x-ray detectors have been developed to separate the x-ray spectrum by energy rather than by wavelength. The energy-dispersive detector can be added to an SEM and has enough sensitivity to show x-ray spectral data at the low-beam currents of SEM. That is, x-ray scanning may be done in an SEM.[44]

1.8. HISTORY OF FIELD-EMISSION MICROSCOPES

In 1897, R. W. Wood described the phenomenon of field emission of electrons, the process of emitting electrons from an extremely small area of cathodic surface in the presence of a strong electric field. In 1936, E.

W. Müller (1911–1977) applied this principle to a negatively charged, very fine tip (< 1-μm radius) of tungsten wire in the high vacuum of a cathode-ray tube. In this, his field-*electron* microscope, Müller obtained a pattern on the fluorescent screen that represented the array of atoms, including impurities or other atoms foreign to the elemental metal of the emitting tip.[31]

In 1950, Müller charged the acicular tip positively, introduced helium into an extremely high vacuum, and formed He$^+$ ions at the tip. Some of the atoms hopped off the tip and activated the fluorescent screen to form a pattern typical of the field-*ion* microscope.[31]

1.9. SCANNING ACOUSTIC MICROSCOPY

There are two types of acoustical microscopes; both are based on the idea of the Russian scientist S. Sokolov, who proposed in 1936 that short wavelengths of *ultrasonic* energy be used instead of light to look directly inside an opaque specimen. This idea was not put into actual practice until the 1970s, when the manufacturing of working models was begun (see Chapter 16).[46]

2

Definitions, Attributes Contributing to Visibility, and Principles

2.1. DEFINITIONS AND ATTRIBUTES CONTRIBUTING TO VISIBILITY

Microscopy is the science of the interpretive uses and applications of microscopes.[1-3] The key word is *interpretive* (see Figure 1.4).

A distinction is made between the terms *microscopical* and *microscopic*: *Microscopical* is the adjective pertaining to a microscope or the use of the microscope.[1,3] *Microscopic* is the adjective pertaining to a very small object or to its fine structure. A *microscopic* particle requires *microscopical* examination to be adequately visible.[1,3]

Microscopy, like macroscopy, is based directly on sight, which is by far our most important sense. Indeed, of all we know, we learn most by sight alone, so that it behooves us to look at the many *attributes contributing to visibility*, both in kind and extent.[2] They are:

1. *Thought and attitude:* James Thurber could not see through a microscope because he thought he could not.[4]
2. *Memory and experience:* Microscopical images are pictorial, and so are easily remembered and easily recorded photographically.
3. *Imagination:* Albert Einstein once said, in effect, that imagination is more important than knowledge.[5]
4. *Resolving power:* The ability to reveal detail in an object by means of the eye, microscope, camera, or photograph.[1,3]
5. *Resolution:* The minimum distance between points or parts of the object as seen by the eye or in a photograph.[1,3]
6. *Contrast:* In black and white, or in color, the ability to distinguish between parts of the object.[1,3]

15

7.   *Correction for aberrations* (image degradation): Kind and extent: refractive, chromatic, or astigmatic.[3]

8.   *Cleanliness and orderliness:* The quality or state of being free of dirt, contamination, or admixture.[3]

9.   *Depth of focus:* The thickness of the image space that is simultaneously in focus.[3]

10.   *Focus:* What you choose to do with the rays to your eye, camera, or microscope to make the particular image.[3]

11.   *Illumination:* The act of supplying radiation to the object.[3] Few objects are self-luminous.

12.   *Radiation:* Emission or transfer of energy as waves in the visible, ultraviolet, infrared, electron, or x-ray ranges.[3]

13.   *Anisotropy:* Having properties that vary with changing direction through the specimen.[3]

14.   *Magnification:* A ratio of the size of an image to the size of the corresponding object.[1]

15.   *Field (of view):* The visible portion of the object.[3]

16.   *Antiglare devices:* Ways of eliminating or limiting rays that do not contribute to the formation of an image.[3]

17.   *Cues to depth (3-D):* Shadows, optical sectioning (differential focus), stereoscopy.[3]

18.   *Depth of field:* The thickness of the object space that is simultaneously in acceptable focus.[3]

19.   *Working distance,* The depth of space between the surface of the specimen and the front surface of the objective lens (or the unaided eye).[3]

20.   *Structure of specimen:* The way a specimen is constructed from parts such as atoms, radicals, or molecules in configuration and conformation in the liquid, glassy, rubbery, or crystalline state and in the phases of a system.[3]

21.   *Morphology of specimen:* The size and shape of parts or particles, topography of a surface, habit of a crystal, or distribution of phases of a system.[3]

22.   *Information about specimen:* Indication of what to look for and how to interpret its image.

23.   *Experimentation:* The testing of a hypothesis, based on interpretations of observations or other experience with the sample.[3]

24.   *Behavior of specimen:* Changes in properties with time, temperature, humidity, radiation, illumination, or other environmental factors.[3]

25.   *Preparation of specimen:* Whatever is necessary for the microscopical examination. Since any preparation changes the specimen in

some way and to some extent, the process should be thoroughly understood in the interpretation of the microscopical image.

26. *Photomicrography:* Separate aid to visibility.

2.2. PRINCIPLES IN TERMS OF ATTRIBUTES

2.2.1. Thought, Memory, and Imagination

The process of vision consists of three components: physical, physiological, and psychological.[6] While rays, waves, and photons are physical and the functions of the eye[7,8] are physiological, the functions of the brain are psychological: thought, memory, and imagination.

2.2.2. Resolving Power

Resolving power of a lens, including that of the eye, theoretically depends upon the angular aperture AA, the angle between the most divergent rays that can pass through the lens to form the image of the object.[1] Figure 2.1 (not drawn to scale) indicates that the angular aperture of the eye's lens is very small (about 10 minutes of arc).

Figure 2.2a illustrates that the angular aperture, AA, of a microscopical lens is purposely much larger than that of the eye. In Figure 2.2b, point P is covered with a glass cover slip of refractive index n, so the angle a in glass is somewhat smaller than the angle α in air.[1] The relationship is: $n \sin a = \sin \alpha$. In Figure 2.2c, the space between the cover glass and the front surface of the lens has been filled with immersion "oil," which is an inert liquid of the same refractive index as that of the coverglass. In such a system the objective is of the homogeneous oil-immersion type. Its effective angular aperture (not shown) has been increased by the factor n. The term $n \sin \alpha$ is called the numerical aperture,[1] NA, where n is the refractive index of the least-refracting

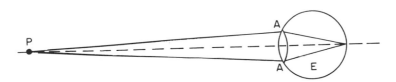

FIGURE 2.1.   The point P subtends an angular aperture AA to the eye (E). The angle is very small compared with the AA of microscopical lenses[9] (Figure 2.2 and Table 2.2). Courtesy of Microscope Publications, Ltd.

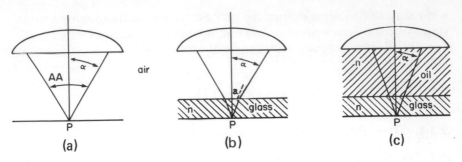

FIGURE 2.2. Effective angular aperture of (a) dry, (b) covered, and (c) homogeneously immersed specimens.[9] Courtesy of Microscope Publications, Ltd.

medium in the path.[9,10] NA is the symbol for the theoretical *resolving power* of microscopical objectives. It is the most important single factor that you pay for in an objective.

2.2.3. Resolution

Resolution is usually expressed as the minimum distance, d, between two lines or points in an object that appear separated in the image.[1] Actually, resolution is the revelation of the two lines or points on two different receptors, separated by at least one receptor on the retina of the eye.[10] It is said customarily that the smaller the distance, d, the greater the resolution.[3]

Abbe theorized from the law of diffraction that $d = \lambda/n \sin \alpha = \lambda/NA$, where λ is the wavelength of the radiation. Consequently, the value of d is smaller (resolution is greater) the shorter the wavelength.

2.2.4. Contrast Perception

Contrast perception is the ability to differentiate various components of the structure of the object by different intensity levels in the image.[3] Unless there is contrast between the spaces and the points, they are not visible even though resolved. In visible-light microscopy there are two kinds of contrast: so-called black-and-white contrast and color contrast. Even in the infrared or ultraviolet regions, if they are multichromatic, color contrast may be obtained by means of electronic image converters.

Contrast by transmitted light depends upon absorption (color), refraction (Becke band or shadows), and thickness of specimen. The purest color contrast is obtained by eliminating the refraction image. This is done by mounting the entire specimen, or a representative portion thereof, in a liquid of similar refractive index. On the other hand, maximum contrast by refraction is obtained by simply mounting in air (which has a very low refractive index). In this, the dry condition, the specimen appears in its darkest state because it reflects and refracts the greatest proportion of light away from itself. Reciprocally, there is the least transparency and resolution of details of the specimen. Other considerations for a mounting medium are viscosity, reactivity, and hardenability (for permanent mounts).[10]

The investigator is fortunate if transparent portions of an object, as received, are separated by opaque ones, or vice versa. He is even more fortunate if the portions of one type are of a *color* different from those of another type. If not, he must know how to *treat* the specimen to produce relative differences in opacity or color. For example, osmium tetroxide (OsO_4), added to certain tissues or synthetic phases, is reduced differentially to opaque black osmium, thereby differentiating the structures. Iodine dissolved in KI, $ZnCl_2$, and H_2O (as Herzberg's stain) colors different cellulosic fibers (and other tissues) differently, according to origins and treatments (for paper, textiles, fillers, etc.). Iodine-based reagents also color other polymeric carbohydrates, and so are used to detect starches and to estimate their degree of dextrinization.[10]

2.2.5. Refractive Aberrations

Refractive aberrations are degradations of image by a transparent, solid lens with either two spherical surfaces or one spherical surface and one plane surface. These are the only two kinds of surfaces which are obtainable with the manufacturing tolerance of one-quarter of a wavelength of light, or about 0.135 μm.* There are five kinds of refractive aberrations: spherical, coma, astigmatism, curvature of field, and distortion.[9]

By far the most important refractive aberration is the one called spherical because it occurs even with a single truly spherical surface (illuminated with parallel monochromatic rays), as shown in Figure 2.3.[9] Rays nearer to the center of the lens travel farther through the refractive solid and are focused at a greater distance from the lens. Of course, cutting out the peripheral rays, regardless of how it is done, reduces the "circle of confusion," but as discussed in Section 2.2.2, the resolving power is thereby reduced.

The correction for spherical aberration, as illustrated in Figure 2.4, is made by combining a double convex lens with a concave–convex lens of proper shape and refractive index to add spherical aberration of equal

---

* μm stands for micrometer, which is equivalent to 10^{-6} meter; formerly the term micron (μ) was used.

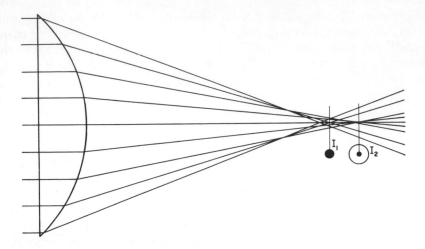

FIGURE 2.3. Spherical aberration.[9] Courtesy of Microscope Publications, Ltd.

degree but opposite direction. Since the rate at which the distances decrease through the first lens is not equal to the rate at which the distances increase through the second lens, the correction is not perfect, but it can be acceptable. So much for rays parallel to or on the axis of the lens.

For rays from any point off the axis on an object of finite size, there is an aberration called coma, because the image of a sharp point (O in Figure 2.5) is shaped like a blurred comet. The condition arises from

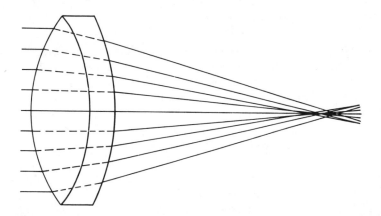

FIGURE 2.4. Partial correction of spherical aberration.[9] Courtesy of Microscope Publications, Ltd.

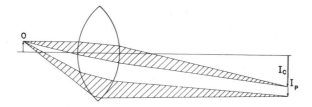

FIGURE 2.5. Coma. O stands for object; I_c is the image formed by rays having passed through the central zone of a lens system; I_p is the image of larger size, formed by rays passing through the peripheral zone of a lens system.[9] Courtesy of Microscope Publications, Ltd.

greater magnification by rays going through the periphery of the lens than by rays going through the central part. The correction for coma is much like that for spherical aberration: The ratio of the sines of the angles which any two rays make with the axis on one side of a lens is made the same as the ratio of the sines on the other side.[10]

The corrections for spherical aberration and for coma assume a certain object distance (Figure 2.6a).[9] If the object is farther away than designed, the lens is undercorrected (Figure 2.6b); if nearer, the lens is overcorrected (Figure 2.6c).[9] It follows that for high-powered, well-corrected light-microscopical objectives the designer assumes a definite mechanical tube length: for transmitted light, 160 mm in America, 170 mm in Germany, or as marked.* The designer also assumes a thickness

---

* Objectives corrected for infinity are not affected by variations in tube length.[9,10]

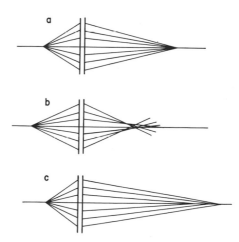

FIGURE 2.6. (a) Correction of spherical aberration. (b) Undercorrection for longer object distance and overcorrection for shorter object distance.[9] Courtesy of Microscope Publications, Ltd.

for the cover glass, if one is intended for high, dry objectives (see Figure 2.2). In America, the thickness of the cover glass is standardized by the American Society for Testing and Materials as 0.17–0.19 mm ("number $1\frac{1}{2}$"), unless otherwise specified by the manufacturer of the objective.[11] For dry objectives of high NA a deviation of as little as 0.01 mm from the designed thickness may be important; hence premium dry objectives are provided with a cover-glass correction collar to provide final focusing for maximum simultaneous sharpness and contrast.[9] The thickness of the cover glass is of no consequence when the proper oil is used with homogeneous immersion objectives for transmitted light. With reflected light the objectives are usually designed for use without a cover glass.

Astigmatism is an aberration occurring in a lens that is not a figure of rotation around its axis. Since microscopical lenses are ground and polished to perfectly spherical surfaces by rotation, astigmatism is not a problem in light microscopy. Astigmatism in the human eye requires correction by prescribed eyeglasses which should never be removed when the student or researcher uses a microscope. Eyeglasses prescribed merely for near-sightedness (myopia) or for far-sightedness (hyperopia) need not be worn by the microscopist because the act of focusing the microscope puts the visual image onto the retina just as well as the prescribed eyeglasses. Moreover, eyeglasses used solely for myopia or hyperopia should *not* be worn, lest they be scratched by eyepieces or require the eyes to be so far above the eyepiece as to reduce the field of view.

Astigmatism *is* just as much a problem with electron-microscopical lenses as it is with closely related television lenses, when round test patterns are not pictured round. Correction for astigmatism in electronic lenses is dicussed in Chapters 12 and 13.

Curvature of field is inherent in solid lenses with spherical surfaces, as illustrated in Figure 2.7. Unlike spherical aberration wherein no part of the image is in sharp focus, within its limits a part of the curved field *is* in sharp focus. Fortunately, the eye has such depth of field in simultaneous focus that curvature of field is not usually a visual problem. In

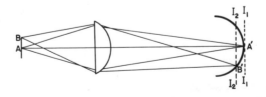

FIGURE 2.7.   Curvature of the field.[9] Courtesy of Microscope Publications, Ltd.

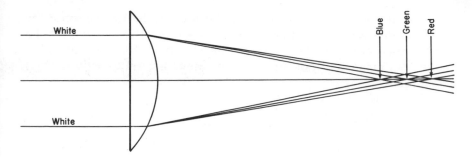

FIGURE 2.8. Chromatic aberration.[9] Courtesy of Microscope Publications, Ltd.

photomicrography on a flat plate or film, however, the image from the objective must be made flat, as discussed in Chapter 8.

Distortion is not usually a problem with monobjective light microscopes, but with stereobinocular microscopes it might image a square as a "barrel" or "pincushion."[9]

Chromatic aberration results when rays of white polychromatic light pass through a lens, primarily because blue light has a shorter wavelength than green light, which has a shorter wavelength than red light. Even so, a central ray passing through the lens is not dispersed into colors, since it is not bent. However, peripheral rays are bent, and the angle is sharper the shorter the waves, as indicated in Figure 2.8.[9] The angle also depends upon the dispersive characteristics of the lens material (differences in refractive index with wavelength). Thus chromatic correction can be made simultaneously with the correction for spherical aberration (Figure 2.4) by choosing the proper dispersion characteristics for each component of the lens system. Chromatic correction for two extreme wavelengths and spherical correction for one intermediate wavelength results in an *achromat* (achromatic lens). Achromats are usually satisfactory for visual work, while *apochromats* are preferred for detection of faint color, for fidelity in color photomicrography, or for focusing in visible light to photograph in ultraviolet light. Apochromats, usually marked as such, are corrected chromatically for three widely different wavelengths and spherically for two.[10] They are corrected extremely well for all aberrations except curvature of field, and that is done by the manufacturer with his particular compensating eyepiece. Fluorite objectives are very similar to achromats, except that clear mineral fluorite (crystalline CaF_2) is used in one or more components of the objective because it has a low refractive index and low chromatic dispersion. Such objectives are intermediate in performance between achromats and apochromats.[9]

Now it should be apparent that besides numerical aperture the kinds and degrees of correction for aberrations contribute to the cost of a microscopical objective. Yet whatever the price paid, the full potential in quality of image depends upon many other attributes to visibility, which follow.

2.2.6. Cleanliness and Orderliness

The microscopist must exercise extreme cleanliness in his work. In light microscopy this means he must avoid all oils and solvents and assure himself that there are no fingerprints on lenses, slides, cover glasses, stages, or mirrors. The inside of the microscope, be it a light microscope or an electron microscope, must be free from all dust and dirt. The specimen itself should be clean and well prepared. The workroom, microscopical desk, instruments, notebook, drawings, and photographs should be clean and orderly.

A word of caution about cleaning the microscope is not amiss. All optical parts should be cleaned only with lens tissue, soft brushes, and lintless swabs. Water is best for removing fingerprints from glass; alcohol for fingerprints from photographs. To remove oil from objectives use a solvent no stronger than xylene, lest cement be loosened. Be extremely careful about abrasive dust such as from cutoff, grinding and polishing machinery, or from abrasive papers, pastes, soil, ceramics, concrete, and other industrial dusts. Be careful with chemical reagents, particularly acids, alkalies, and fluorides. If something as corrosive as these reagents must be used, attach, with immersion oil, a cover glass to the objective you need, use reagent as necessary, and put the rest away! Always be careful not to focus toward a specimen covered with or resting on glass. Broken glass scratches the lenses because optical glasses are softer.

2.2.7. Depth of Focus

Depth of focus in the image space[1,3] refers to the other side of a particular lens with respect to depth of field in the object, as shown in Figure 2.9.[9] Although the terms depth of focus and depth of field may be considered synonymous,[12] they are discussed separately here, since depth of field is objective in meaning while depth of focus is subjective with respect to the microscopist. If he is working visually, it may be an advantage to have an image equal to the great depth of focus of the human eye with its low numerical aperture, NA, 0.002.[10] With a deeper image (Figure 2.9) the observer does not have to twiddle with the focus-

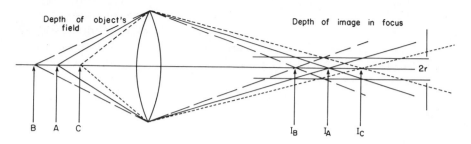

FIGURE 2.9. Depth of field[9] and depth of focus. Courtesy of Microscope Publications, Ltd.

ing wheels as often, and he may already have his hands full. In such cases he would choose an objective of low NA.

On the other hand, great depth of focus is a disadvantage in optical sectioning when studying the spatial arrangements of structures.[10] In such studies a very thin depth of field is required to observe the sharp changes in the object's contours with very small changes in focus. Hence an objective of high NA is selected.

2.2.8. Focus

In visual microscopy the eye (Figure 2.10) should be focused at infinity so that the ciliary muscles (M) controlling the thickness of the eye's lens (L) are relaxed, in order to avoid eyestrain and headache. That is, imagine that the object is on the moon rather than under your nose. If your microscopical image goes out of focus, refocus the *objective* rather than "accommodate" with the ciliary muscles.[10]

Objectively, focusing means turning mechanical wheels or electronic knobs to move the object or objective, or to change the optics

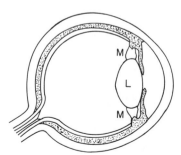

FIGURE 2.10. Diagrammatic section through eyeball. M, ciliary muscles; L, lens, elastic.

somehow. Subjectively, focusing is a personal choice of image by the direct observer, the photomicrographer, or the "armchair" microscopist sorting out data. Focusing is not always choosing the sharpest image; sometimes it is the process of over- or undershooting to gain contrast, resolution, or some other attribute to visibility. Whatever else it may be, focusing boils down to each individual's personal choice in interpretation of images.

2.2.9. Illumination

Illumination is very important and has to do chiefly with the *condenser*. Figure 2.11 illustrates the kinds of illumination: symmetrical, axial, bright-field, and dark-field for transmitted and for reflected light,

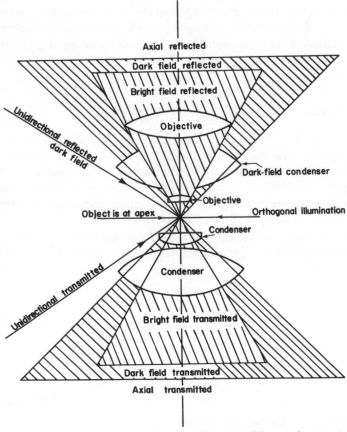

FIGURE 2.11.   Principal types of illumination. Directional beams lie on or within a principal cone shown above.[14]

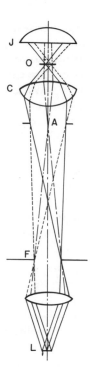

FIGURE 2.12.   Illumination system for transmitted light (Köhler illumination).[9] Courtesy of Microscope Publications, Ltd.

with grazing and intermediate unidirectional illumination in between.[13] For symmetrical reflected illumination the objective serves as its own condenser.

The substage condenser can be used in one of two ways. In Nelsonian ("critical") illumination, parallel incident rays from the light source (sun, lamp filament, or other objects such as a linear or areal scale or a picture) are brought to focus in the plane of the object, as shown in Figure 2.11.[14] If a lamp is used, an auxiliary condenser (usually in the lamp housing) serves to render its emergent rays approximately parallel. The chief limitation is that the size of the *illuminated field* cannot conveniently be made to equal the objective's field. Glare (nonimage-forming light) may result, reducing contrast. Another limitation occurs if the image of the lamp's filament does not produce an evenly lighted field. Adjustments such as throwing the filament out of focus, or sanding the glass envelope, spoil the "critical" aspect.

In Köhler's illumination (Figure 2.12), the source of illumination L is focused in the plane of the condenser's diaphragm A. The substage condenser C is moved to bring the image of the lamp's iris diaphragm into the plane of the object O. Then the lamp's diaphragm F is opened just enough to fill the objective's field of the object O. Another advantage is that the lamp's filament is out of focus, so that the field is evenly illuminated. The field stop F also enables the observer to center the optical axis to coincide with that of objective J. The substages of microscopes are generally equipped with centering devices for the condenser system. When both F and O are in focus, the centering screws should be used to center the field-stop image with the field of view at O.[9]

According to Abbe, the formula for the resolution of a periodic structure of units separated by distance d and illuminated by rays parallel to the optic axis y is $d = \lambda/n \sin \alpha$, where α is the angle between the axis and the diffracted rays of the first order (Figure 2.13). λ is the wavelength of monochromatic light or of the longest of mixed wavelengths, and n is the refractive index[1] of the least-refractive medium between the object and the first lens of the objective.[10]

Abbe reasoned further that resolution is theoretically doubled if the incident rays (O, Figure 2.13) are rendered so oblique that the angle of incidence (i) is equal to α, the angle of the two diffracted rays of the first order; then *both* the *undiffracted* ray O and *one* of the two *diffracted* rays enter the objective. Therefore d, the distance between units barely resolved, is theoretically cut in half. That is, $d = \frac{1}{2}(\lambda/n \sin \alpha) = \lambda/2NA$.[10]

The illuminating condition of maximum obliquity requires that the numerical aperture of the condenser be equal to that of the objective (Figure 2.13), but this requirement is seldom met because the contrast is so poor. Therefore, the iris diaphragm of the condenser (Figure 2.14, left) is usually stopped down to give a solid cone of light equal to three-quarters of the objective cone, or less.[13,15] The diaphragm is closed to obtain sufficient contrast between the units and spaces, and to obtain sufficient depth of focus if the object has depth of field. The practical equation for resolving power then becomes $d = \lambda/kNA$, where k is a constant that varies from 2 to 1, respectively, when the condenser diaphragm is wide open or closed to a pinhole. If the illuminating cone is two-thirds of the objective cone, k becomes $1.00 + 0.67 = 1.67$. Then $d = \lambda/1.67NA = 0.60/NA$.

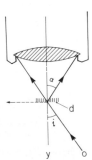

FIGURE 2.13. The optimum value of oblique illumination.[14]

FIGURE 2.14.   (Left) Iris diaphragm (variable annular stop). (Right) Traviss type of expandable central stop.[13,15]

The equation is almost exactly like Lord Rayleigh's (John W. Strutt, 1842–1919), which he arrived at while considering the image of a self-luminous body such as that of a fluorescing particle. He found the image to be a diffraction (Airy) disk. The images of two self-luminous bodies will touch when the distance d between them is given by $1.22/2NA = 0.61/NA$.[10] These two different routes of reasoning have brought us to the same theoretical limit of resolution, but Abbe's theory is preferred here because it deals fundamentally with the role of the condenser, especially in *bright-field illumination*, which thus far has been under consideration.

However, in *dark-field illumination* the particles or parts of a specimen appear to be self-luminous, so that here Lord Rayleigh's theory is more relevant than Abbe's. As indicated in Figure 2.11, dark-field transmitted illumination is obtained from a condenser of greater angular aperture than that of the objective, by centrally stopping the condenser so that no light can enter the objective directly. Then the field is dark, but light getting around the central stop, and thereby passing through the condenser as a hollow cone, can be scattered by parts of the specimen into the objective. These parts appear bright (self-luminous) against the dark field. For example, particles are visible in high contrast, but like stars in the sky they appear as (Airy) disks of light, without sharp outlines.

The central stop can be given the proper diameter by means of an expandable (Traviss) type of device (Figure 2.14), which will not fit into every manufacturer's substage, since provision for it has not been standardized. The handy person can probably improvise a fitting. A specific manufacturer may provide simple central diaphragms of fixed stop sizes commensurate with his own objectives and condenser. To date these are not standardized either, but again the skilled microscopist could modify a misfit or start with solid-sheet material and improvise entirely.

Dark-field illumination by transmitted light for objectives of very high NA is discussed in Section 3.5.11 in Chapter 3. Dark-field illumination by reflected light is considered in Chapter 4, and dark-field illumination in a transmission-electron microscope (TEM) is discussed in Chapter 12.

The intensity of illumination should be such that it is not much greater than that of the room. The illumination should never be so bright as to cause squinting or complete closing of the eye's iris diaphragm (Figure 2.10), lest the controlling muscles become tired and painful or even paralyzed (called "microscope eye"). The intensity should be controlled at the source of illumination, never by closing the diaphragm of the condenser or that of the lamp; they are for entirely different purposes. For an incandescent lamp, a variable transformer to control voltage is the best way to control intensity. Since reducing the voltage will change the color temperature toward the red end of the spectrum, a bluish "daylight" filter is inserted whenever color fidelity is important.

2.2.10.   Radiation

Daylight *radiation* is a spectrum of wavelengths varying in color, as shown in Table 2.1.[16]

TABLE 2.1
Colors and Their Corresponding Wavelengths

Color	Wavelength (nm)[a]
ultraviolet	200–400 (invisible)
violet	400–455 (visible)
blue	455–492 (visible)
green	492–550 (visible)
yellow	550–647 (visible)
orange	588–647 (visible)
red	647–700 (visible)
infrared	700–3000 (broad invisible range)

[a] nm stands internationally for nanometer, a millionth of a millimeter (10^{-9} meter). A nanometer was formerly called a millimicron (equal to 10 Å), a unit which is internationally obsolete.

Color is visually important in microscopy for many reasons. Specimens may display color absorption, transmitting the remaining color portion of the spectrum.[10] A specimen mounted in a medium of similar refractive index will still be visible if it displays a color image. A specimen may be very close in refractive index n to that of its mounting medium for one wavelength, but quite different in n for another (usually complementary) wavelength.* The result is a colored boundary like the Christiansen effect (dispersion staining),[10,17] which is structural color that depends on optical phenomena of reflection, refraction, and diffraction in a fine structural system of phases.[10] Thin films often manifest interference colors. Polarization colors are also interference phenomena. All these kinds of colors (and more) make variation in color an important attribute contributing to visibility.

It is important that the optics of the light microscope be corrected for chromatic aberrations lest such artifacts be misinterpreted. If a microscopical objective is to be used photographically in the ultraviolet region (200–400 nm), it should be corrected chromatically for a wavelength in this region (e.g., 253.7 nm, a spectral line of the mercury vapor lamp), while the visible range is filtered out. For the purpose of focusing, the UV objective should be corrected also for a visible color [e.g., the green line (508 nm) of the mercury spectrum] while the green filter and a no-UV filter are used. A UV objective is useful in studying ultraviolet absorption by a molecule or radical, such as absorption by melamine in a melamine–formaldehyde resin.[2]

Another important aspect of radiation as an attribute to visibility is the role of wavelength λ in determining the theoretical limit of resolution, expressed as $d = 0.6\lambda/NA$. Simply using UV light (e.g., 253.7 nm), instead of green light (e.g., 508 nm), cuts d in half and doubles the theoretical resolution. The use of UV solely to increase resolving power has become obsolete now that electron microscopes utilize wavelengths thousands of times shorter. The rapid advance in electron microscopy stems from the discovery that beam electrons obey the same laws as rays of light. (See Chapters 12, 13, and 14.)

This brings us to the *sources* of radiation.[18] In light microscopy the incandescent tungsten filament remains the most important for visual work. For low-power microscopes with relatively large fields, fluorescent lights (especially circular ones) are gaining in availability and demand.

For cinephotomicrography, phase-amplitude contrast, interferometry, schlieren microscopy, and some work between crossed po-

* Such variation of refractive index with wavelength is called *dispersion*. See reference 1 and discussion of chromatic aberration, Section 2.2.5.

lars, the xenon arc and the quartz–halogen lamp are taking precedence over the older types such as the carbon arc and the mercury vapor lamp.[18]

2.2.11. Anisotropy

Anisotropy may originate in a structure in one or more of five manifestations of influence by polarized light:

1.   *Molecular* anisotropy resulting from the orientation of dipoles, usually in long or flat molecules[19];
2.   *Birefringence*[1] in *units* of all *anisometric* crystals;
3.   *Polarization patterns,* such as the permanent cross seen in *spherulites*[3] like starch grains;
4.   *Flowed, streamed,* or *grown* masses of long or flat units (not necessarily crystalline) in a medium of *different refractive index*[10,19,20];
5.   Regions of local *strain anisotropy* (photoelastic effect) in a normally isotropic substance such as inorganic or organic glass *under stress*.[2]

Many materials owe their anisotropy to two or more of the above-mentioned origins. Cotton fibers are a good example, as illustrated in Figure 2.15.

All fibers except inorganic glass ones manifest eight independently determinative properties, based on their anisotropy. Thereby both structure and morphology are revealed[2] (see Chapter 6).

Crystals reveal even more optical properties than fibers, based on many variations in crystallographic symmetry, as discussed in Chapter 7.

Liquid crystals are in a class by themselves with regard to their many optical properties based on anisotropic properties.[19] These will be discussed in Chapter 7.

2.2.12. Magnification

Magnification is incidental to the resolving power (NA) of the objective and that of the eye. The approximate relationship is[10]

$$\frac{\text{maximum NA (microscope)}}{\text{minimum NA (eye)}} = \frac{1.40}{0.002} = 700\times$$

which can also be stated as

$$\frac{\text{limit of resolution (eye)}}{\text{limit of resolution (microscope)}} = \frac{0.15 \text{ mm}}{0.0002 \text{ mm}} = 750\times$$

50 μm

FIGURE 2.15. Cotton fibers, mounted in mineral oil and shown between partially crossed polars, showing the convolution of the fibers resulting from the complex structure. Taken by Mrs. G. Berry.

The theoretical optimum magnification for light microscopy therefore is about 750×. Useful magnification in practice may be somewhat higher, in order to assure that adjacent points in the object are imaged on separated receptors (rods and cones)[21] on the retina of the eye (Figure 2.10). By "separated" receptors we mean that the two nerve-endings receiving images of different points of the object are not adjacent. Instead, there must be at least one other nerve-ending between the two. The necessary magnification may be as high as 1000 times the NA of the objective (Table 2.2). Usually any magnification above 1000NA is empty, that is, the additional magnification reveals no more detail of the object.[1,10]

The chief function of the eyepiece is to increase the incidental magnification of the objective to a total magnification that is useful to the eye:

$$\text{mag. (objective)} \times \text{mag. (eyepiece)} = 1000 \times \text{NA (objective)}^{[10]}$$

TABLE 2.2
Theoretical Resolving Powers of Light Microscopes

Focal length (mm)	Numerical aperature, NA	Maximum useful magnification	Approximate theoretical limit of resolution, d^a		Depth of field (μm)
			Axial (μm)	Oblique (μm)	
250	0.002	eye alone, 1×	150	–	–
25	0.10	hand lens, 10×	10	–	42
32	0.10	compound microscope, 100×	5	2.5	25
16	0.25	compound microscope, 250×	2	1	3.8
8	0.50	compound microscope, 500×	1	0.5	0.86
4	0.95	limit, air-immersion objective, 1000×	0.52	0.26	0.083
3	1.38	oil immersion (visible, 500 nm), 1500×	0.36	0.18	–
3	1.38	oil immersion (UV, 270 nm), 2000×	0.20	0.09	0.04

^{a} Based on Abbe's theory: $d = \lambda/(1 \text{ to } 2\,NA)$ (disregards aberrations). This is about *maximum* obtainable resolution and assumes adequate contrast. Any reduction of aperture of objective or condenser increases d. Photography has its own limitations.

The eyepiece magnifies the real image formed by the objective inside the eyepiece, in the plane of its fixed diaphragm. The diaphragm not only frames the image, but also receives any reticule such as a micrometer scale that you wish to superimpose on the image of the object.

The two principal kinds of eyepieces are the Huygens and the Ramsden (see Chapter 1). Both designs employ two planoconvex lenses, as shown in Figure 2.16. In the Huygens type (Figure 2.16a), both plane surfaces face toward the eye. Its field lens (F) bends the rays from the objective to form a real image of the object in the plane of the diaphragm (D). The rays from the eye lens (E) cross over at the "eye point" (E_1-E_2) to form a small bright spot. From here the rays enter the eye and form an image on the retina. In the Ramsden design (Figure 2.16b), the field diaphragm D is in front of the field lens F, whose plane surface is faced outward.[9] This characteristic makes the Ramsden type easily recognizable. Correction for aberration may be achieved as a doublet or triplet in the field-lens system. The ease and convenience with which this is done makes the Ramsden design the preferred type for eyepieces of magnification greater than about 12×.[10] The Huygens type is preferred for eyepieces of magnification less than 12×. Both types are intended for achromatic objectives. Apochromatic objectives usually have some residual curvature of field which is corrected by means of specific compensating eyepieces or by some generalized flat-field and wide-field eyepieces.

All eyepieces have eye "points" like the crossover spots, shown as E_1-E_2 for both Huygens and Ramsden eyepieces in Figure 2.16. The position of the eye point can be seen as the brightest, smallest spot when a small piece of ground glass or translucent sheet is moved slightly toward and away from the eyepiece. If eyeglasses are prescribed to correct for myopia only or for hyperopia only, they may be removed so that the naked eye may be placed at the eye point. Focusing the microscope takes care of the near- or far-sightedness, but if the eyeglasses correct for astigmatism, they should be worn at the eyepiece. When glasses are worn, the eye may then be so far above the eye point that the field of view is reduced. For this reason some manufacturers have produced eyepieces with especially high eye points.[9]

Precise designation of magnification is generally important on photomicrographs and other projections. If a ground glass can be used in place of a photographic film, the image of a standard-stage micrometer can be measured on the ground glass or projection screen. The magnification can then be calculated as a simple ratio. The magnification of a contact print will be the same. The magnification of an enlargement would include the magnification of the enlarger, obtained by projecting the image of a transparent standard onto the easel. When using roll film, one frame is devoted to recording the negative image of a stage micrometer. The positive image is measured subsequently and compared with the original micrometer, giving the precise magnification factor.

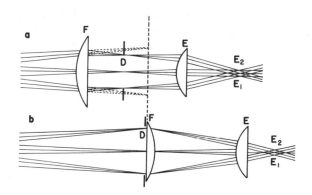

FIGURE 2.16.   Huygens (a) and Ramsden (b) oculars.[9] Courtesy of Microscope Publications, Ltd.

2.2.13. Field of View

Field of view originates with the *objective*. It is first reduced in area by the diameter of the microscope's tube, or by the diaphragms in it. Finally, the field of view is bounded by the diaphragm in the focal plane of the eyepiece. Microscopical tubes and eyepieces are of two standard sizes. The wider tube accommodates especially widefield eyepieces.

2.2.14. Antiglare Devices

Antiglare devices include the coatings (usually bluish) of a low-refracting substance (e.g., magnesium fluoride) on the air surfaces of lenses. The blue is the interference color of so thin a coating. The purpose of the coating is to reduce the marked difference in refractive index between glass and air, and so reduce the intensity of the reflection from the polished glass surface. Such reflected light is undesirable glare. Polarization of the incident light also cuts down glare, but for a different reason: Reflected or scattered light is partially polarized, and therefore it can be eliminated by a properly oriented polar. Köhler's illumination (see Section 2.2.9) also reduces glare.

2.2.15. Cues to Depth

Three-dimensional cues are not at all as numerous in microscopy as in macroscopy. Shadows, produced with oblique illumination by masking part of the aperture of the condenser, are the best cues given by transmitted illumination. Also, the Becke line, a diffraction line produced by slightly over- or underfocusing in a proper mounting liquid, is useful to show depth. If there are curved or oblique faces on the object, they also can produce relevant shading.

By reflected light, unidirectional or semisymmetrical illumination produces revealing shadows if the object has sufficient depth, either as it is received or relief-polished or etched.

Stereoimages are readily obtained with any stereoscopic microscope[22] by taking a picture through each of the two eyepieces. The resulting stereopair is viewed in a stereoscope or by stereoprojection. Stereoimages are not readily obtained with a nonstereoscopic microscope, except by means of a stage which may be tilted the required angle right and left of the microscopical axis.

2.2.16. Working Distance

Working distance, or lack of it, is a problem of objectives for visibility when using bulky accessories such as a hot or cold stage or a micromanipulator. The stereomicroscope has objectives of long working distances, and is good for viewing relatively large objects. Microscopes equipped with a universal stage usually are sold with objectives of long working distances. The objectives carry a standard thread[23] which permits interchangeability with other microscopes. There are also other standard objectives of long working distance for special visibility, such as those used for looking into red-hot ovens.

Working distance is important for ordinary "high-dry" objectives because it determines how thick the object and optimum cover glass (No. $1\frac{1}{2}$, 0.18 mm) may be.[11]

2.2.17. Depth of Field

Depth of field[1,3] relates visibility to the *object*. If parts of an object are to be in simultaneous focus, the depth to be penetrated by transmitted light or reflected light should be no greater than the depth of focus of an objective of particular NA (Table 2.2). This means that a transparent section should be thin enough to use with objectives of high NA. The standard thickness for rock sections is 30 μm. With reflected light a surface must have a depth of field within the same limits.

2.2.18. Structure of the Specimen

Manifestations of *structure*[1,3] which are microscopically visible depend in large measure upon the nature of the object. Abbe's fundamental theory of resolving power *assumes* that the specimen is of microscopic, periodic structure (see Section 2.2.2). A diatom is of periodic structure, but most specimens are not. Even so, the diffraction image of a certain diatom with *hexagonal close-packing* of its pores shows hexagons instead of ellipsoids, each with a slit, unless the resolving power is adequate[13] (Figure 2.17).

Transmitted images are refracted. They are formed by bending (i.e., refracting) the light at surfaces or interfaces at which there is a difference in refractive index, i.e., a difference in velocity of light.[1] The velocity of light in a substance depends upon its structure as well as its composition. If the structure is isometric (equal spacing of units in all directions), the substance is isotropic; that is, the velocity of light is the same, no matter what the direction of propagation. Such a substance has no birefringence. Anisometric structures, however, are anisotropic, and they have their own attributes contributing to visibility (see Section 2.2.11).

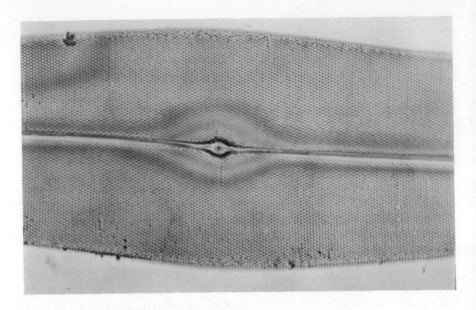

FIGURE 2.17. *Pleurosigma angulatum*, test diatom.[13]

2.2.19. Morphology of the Specimen

The *morphology* (size and shape) of a specimen[1,3] affects its visibility. The smaller the size and the more complex the shape of an object, the greater the required resolving power, contrast, correction of aberrations, and most of the other attributes contributing to visibility to obtain a clear, useful image. Otherwise, transparent particles will look opaque, parallel lines will be fused together, and corners of polyhedra will appear to be rounded. It is said that a regular polygon with x number of sides will be resolved only if its diameter is x times the limit of resolution.[10]

2.2.20. Information about the Specimen

Information about the specimen is often of utmost importance. What are we looking for? Why? Is the sample representative or nonrepresentative (exaggerated)? Is it of a type already on the market? If so, much information on possible solutions to the problem can be gathered from technical literature on that type.

2.2.21. Experimentation

Experimentation with equipment and variation of conditions surrounding a specimen are necessary not only to improve visibility but also to justify interpretation of images. To obtain optimum resolution, contrast, cues to depth, depth of focus, etc., one must experiment with objectives, oculars, condensers, polarized light, different illumination, and all the rest. If one does not have all the equipment, he can improvise.

Of course, keeping good, orderly records is also important for arriving at satisfactory conclusions. Good records and easy retrieval will save a lot of future experimentation regarding subsequent problems, projects, and programs.

2.2.22. Preparation of the Specimen

Every sample should be examined immediately *as received*. If precautions to preserve it are specified, they should be taken. If you clean or fractionate the sample in any way, all fractions including "dirt" should be saved and labeled in case you want them later.

Some *preparations* are routine and need no deliberation: mounting in an inert liquid of chosen refractive index to improve the transparency and visibility, and cutting or otherwise preparing a surface quickly. More drastic and time-consuming preparations should be considered carefully and used on only part of the sample.

2.2.23. Behavior of the Specimen

Behavior means change in the sample with time, temperature, weather (natural or simulated), humidity or dryness, atmosphere (or vacuum), etc. Even during storage or shelf life, samples should be examined often, to look for change and to interpret them in terms of the problem.

2.2.24. Photomicrography

Photography at best is an illustration of the appearance of the specimen under very specific conditions. But photography has its own advantages and limitations of resolution, contrast,[24] color, and emanation sensitivity introduced by grain size or color sensitivity of the photographic film or paper, and by the conditions of its development.[25]

Hence *photomicrography* is a separate large discipline which combines the art of the photographer with the science of microscopy. Frequent reference to photomicrography[25] will appear in all later chapters.

2.3. SUMMARY

Microscopy is a visual science, based on the science of optics, the sense of sight, and the subjective aspects of seeing. There are at least 26 attributes contributing to visibility and seeing: thought and the associated attitude, memory organized into accumulated experience, imagination exercised in interpretation, resolving power of the entire optical system, the resolution of detail thereby obtained, visual contrast, optical aberrations and their corrections, cleanliness of the system, depth of focus, focus of the image, illumination of the specimen, character of the radiation used for illumination, effects of anisotropy, magnification, field of view, reduction of glare, cues to depth, depth of field, working distance, structure of the specimen, morphology of the specimen, information about the specimen, experimentation under the microscope, behavior of the specimen, preparation of the specimen for observation, and photography of the specimen.

The optical principles involved in resolution of structure and resolving power of a lens system were treated first, and then the ways of correcting the system for spherical and chromatic aberrations, for coma, and for curvature of field were considered. The construction and use of achromatic and apochromatic objectives and their special eyepieces were part of this explanation. The distinction between depth of focus and depth of field was then discussed in terms of the act of focusing on different parts of the specimen. Next, the various methods of illumination were considered in terms of optical principles and actual operation under bright-field and dark-field conditions, including the effect of illumination on resolution. If the sample is anisotropic, it provides several additional avenues for obtaining information. The distinction between useful magnification and "empty" magnification was explained, and the limits to useful magnification set by the optical components were considered. Finally, we concluded with brief practical discussions of working distance, manifestations of structure and morphology, preparation of the specimen, and obtaining photographic records of the microscopical image.

Simple and Compound Microscopes

3.1. THE LIMITING RESOLUTION

The limiting resolution d, the distance between two points barely resolved by the human eye, is called visual acuity. It varies directly with the distance D between the object and the eye. In Figure 3.1 the eye's viewing angle V is greatly exaggerated, and the distance D from object to eye is not to scale with respect to d, which separates the barely resolved points P and P. It is seen from the diagram that if the nearer distance D' to the closer points P' and P' is exactly half the distance D, then the barely resolved distance d' between P' and P' is exactly half d.[1] This direct relationship persists as the object is brought closer and closer to the unaided eye, until the object is brought to a certain minimal distance, approximately 250 mm from the normal eye. At this distance of closest vision, the eye can resolve two points which are about 0.15 mm apart.[2] These limits are set by the numerical aperture NA = 0.002 for the eye with its iris diaphragm wide open (Table 2.2). At this setting the eye's lens (Figure 2.10) is as thick as is physiologically possible.

The far-sighted person, whose eyes cannot accommodate for objects much closer than arm's length, requires either eyeglasses prescribed for hyperopia or a magnifying glass to achieve normal visual acuity. Following the same principle, a person with normal eyes can increase his visual acuity by using a magnifying glass. The simple microscope, illustrated in Figure 3.2, follows the same principle. The auxiliary lens bends the rays from the object so that they fall within the eye's viewing angle and are focused on the retina, rather than beyond it.

As shown in Figure 3.2, with a simple microscope a virtual image, P'-P', appears at some imaginary distance, D_i. If this distance is taken to be 250 mm, as in close vision, and the lens or lens system has a focal

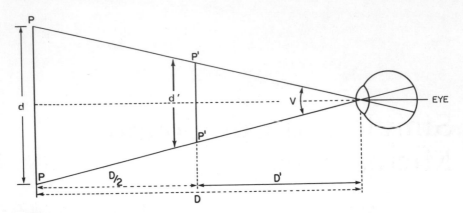

FIGURE 3.1. The viewing angle. The relationship between the distance *d* between barely resolved points *P* and their distance *D* from the eye.[1] Courtesy of Microscope Publications, Ltd.

length of 25 mm, we say that the microscope has a magnification of 10 times, or 10×.

3.2. SIMPLE MICROSCOPE: ONE LENS SYSTEM

Simple microscopes have only one lens system. They are inexpensive, have relatively large fields, and give upright images. Mounted in a

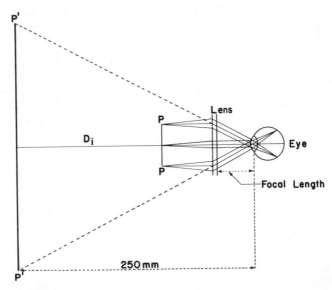

FIGURE 3.2. A simple microscope.[1] Courtesy of Microscope Publications, Ltd.

focusable stand, a simple microscope of large diameter is particularly useful in the dissection of a specimen or in the separation of its parts. Fitted with a micrometer, grid, protractor or other reticle, a simple microscope is also useful for measuring distances, areas, or angles. Some commercial models are fitted with illuminators. The chief limitation of all simple microscopes lies in their relatively low NA, with its corresponding limiting resolution of about 10 μm.

3.3. COMPOUND MICROSCOPE: TWO OR MORE LENS SYSTEMS

A compound microscope is composed of two or more lens systems. There are some compound microscopes which contain only the minimum: an objective plus an eyepiece (Figure 3.3). The objective forms a real image P'-P' in the plane of the eyepiece.[1] The eyepiece is a magnifier which causes the images of two separate points P and P to fall on separated receptors P'' and P'' on the retina.[2]

Figure 3.4 illustrates an uncomplicated compound microscope being used by hand in micrometry. Even smaller models, scarcely larger than a fountain pen, are available. Portability is their main advantage. Low numerical aperture (NA = 0.06 to 0.10) is their main limitation. Moreover, hand microscopes cannot substitute for those with sturdy

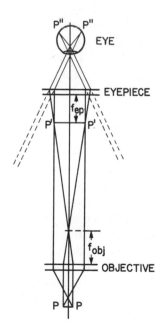

FIGURE 3.3. Light path of two characteristic rays through a compound microscope comprised of only objective and eyepiece.[1] Courtesy of Microscope Publications, Ltd.

FIGURE 3.4. Hand compound microscope being used in micrometry.[6]

stands, so most microscopical work is done with compound microscopes mounted on convenient stands and equipped with accessories necessary for proper illumination and precise focusing.

3.4. STEREO COMPOUND MICROSCOPES

Stereomicroscopes are the most versatile kind of light microscopes to own and use. In almost every microscopical laboratory this type ranks first in variety and frequency of use. There are two kinds: (1) binobjective–binocular and (2) binocular with common main objective (CMO).[3]

The binobjective–binocular stereoscopic microscope (such as the Greenough type) is a classic in microscopy.[2–4] Such an instrument is really two compound microscopes mounted at a slight angle to each other on a single stand. As shown in Figure 3.5, the binocular microscope has an objective and an eyepiece for each eye, mounted at the interpupillary angle of close vision. The natural stereo effect of the binocular system is enhanced by a pair of erecting prisms, adjustable for the individual's interpupillary distance between eyepieces. The paired objectives are slender, compact, and close together, providing easy access to the specimen. Furthermore, aberrations are relatively easy and inexpensive to correct because each of the two beams enters its respective lens *axially*.[3] (See Chapter 2, Figures 2.3 and 2.4.)

The disadvantages of the binobjective–binocular system are that the intermediate images are inclined from the plane of the specimen stage and tilted toward each other. Therefore, only the central portion of the two images are in simultaneous focus. Besides, both images are tilted unless the specimen is tilted for only one image.[3]

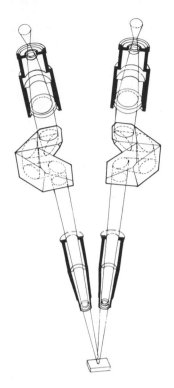

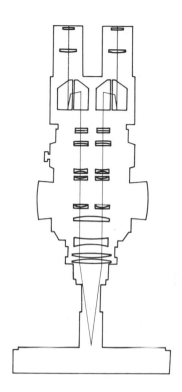

FIGURE 3.5. Greenough stereomicroscope system: A diagram of the beam path.[3] Courtesy of Wild Heerbrugg Instruments, Inc.

FIGURE 3.6. The common main objective stereomicroscope system: A schematic diagram of the beam path.[3] Courtesy of Wild Heerbrugg Instruments, Inc.

The other type of stereomicroscope has only one main objective but has two eyepieces, as shown in Figure 3.6. It may be equipped with erecting prisms, like the Greenough type. The advantage of the CMO is that the intermediate image planes are parallel to the plane of the object without tilting toward each other. Therefore both images are in simultaneous and complete focus. Consequently accessories are easily adapted. However, since the two imaging beams traverse the large common objective obliquely, corrections for aberrations are relatively difficult and expensive to make.[3] (See Figure 2.5.)

Either the monobjective or the binobjective type of stereomicroscope can be equipped with old-style stepwise changes in objectives, or with the newer-style zoom objectives with a lens system of continuously variable magnification.[3,4] One built-in advantage to the stepwise system of objectives of fixed magnification is in micrometry. Once each objective has been calibrated with a given micrometer eyepiece by means of a standard-stage micrometer, the calibrations are fixed.[3]

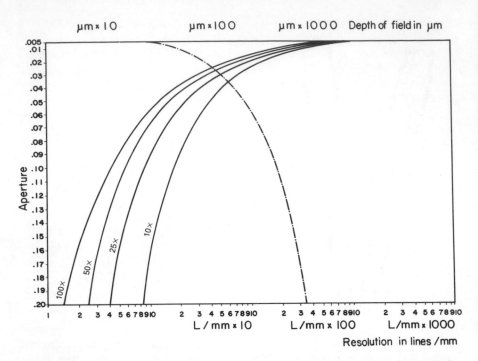

FIGURE 3.7. Resolving power and depth of field in stereomicroscopes. The solid curves, drawn specifically for total magnifications (as shown) obtained with the M5 stereomicroscope, are applicable in principle for other stereomicroscopes made by Wild Heerbrugg. The dash–dotted curve, representing the correlation between numerical aperture and resolving power expressed in lines per millimeter, has general validity for reflected light observation. Courtesy of Wild Heerbrugg Instruments Inc.[3]

The corresponding disadvantage of variable magnification in the zoom objective is obvious, but is partially overcome in a newer design by click-stops on the rotating collar. The manufacturer of the design has published information about its zoom stereomicroscopes relating numerical aperture, resolution, depth of field, and definite magnifications. The data are reproduced as Figure 3.7 and Table 3.1.[3] With reference to Chapter 2, the numerical aperture is independent of the eyepiece's magnification (10× in Table 3.1).

The relatively great depth of field of either type of stereobinocular microscope is a distinct advantage in examining surfaces which have suffered fracture, cracking, corrosion, erosion, weathering, etching, sawing, or abrading. It is also an advantage in examining, sampling, separating, dissecting, selecting, and orienting biological materials, crystals, soils, sands, powders, flours, spots, and specks, particularly on uneven surfaces.

Such an instrument may also be used to examine, at least in a preliminary way, natural, synthetic, or artificial composites and their constituents. The utmost *advantage* of the stereomicroscope is the realistic three-dimensional image of any specimen whose third dimension is within the relatively great depth of focus. A further advantage is the characteristically large field. Ordinary kinds of *illumination* may be employed, and since their effects are known, the image is easy to interpret. Also, the image is erect, and therefore manipulation of the specimen is easy. All these advantages come with a compound microscope, which is comparatively inexpensive.

The chief *limitation* of the stereomicroscope is its relatively low resolving power. There is no substage or condenser, and therefore the kinds and intensities of transmitted illumination are limited. For reflected light, vertical bright-field illumination is very limited in its ability to gain contrast.

The most convenient laboratory stand for the stereomicroscope is one which is separable at the joint between the specimen stage and the base. After separation, the whole microscope may be placed directly onto a large flat object, thus avoiding destructive sampling. In this position, ordinary incident oblique illumination is usually employed. For transmitted light, the base carrying the mirror and the glass window is put into place. Double hand rests are usually supplied for convenience in manipulating the specimen. The complete microscopical

TABLE 3.1

Numerical Apertures of Various Optical Combinations for the M5 Stereomicroscope[a]

Optical combination	Magnification-changer position[b]			
	50×	25×	12×	6×
Main objective alone	0.0800	0.0750	0.0445	0.0223
Main objective with 2.0× attachment objective	0.1650	0.1500	0.890	0.0446
Main objective with 1.5× attachment objective	0.1200	0.1125	0.0667	0.0334
Main objective with 0.5× attachment objective	0.0400	0.0375	0.0223	0.0112
Main objective with 0.3× attachment objective	0.0240	0.0225	0.0134	0.0067

[a] Table courtesy of Wild Heerbrugg Instruments, Inc.[3]

[b] The magnification-changer position indicates total power when 10× eyepieces are used. It should be noted that numerical apertures for each position are independent of the eyepiece power.

head should be removable to go onto other stands, one of which is a heavy stand with a long cross-arm, so that the microscope may be swung over an irregular object or part of a large machine or other assembly.[2,6]

Preparation of the Specimen

For preliminary examination with a stereomicroscope, *no* preparation is desired, much less required. The idea is to look and see exactly what has been received. Anything done to the specimen will change it. Besides, any action will take time which alone may cause the specimen to change.

If the specimen will not stand by itself, stick it in modeling clay, or flatten a bit of it to make a base. If the specimen is smooth enough for its surface to be within the depth of focus of a low-power objective, but is not all in focus with a higher-power objective, rather than smoothing it (and changing the surface) revert to a lower-power objective and choose a higher-power eyepiece.

For the reasons pointed out above, no loss of numerical aperture or resolution results from this practice. To expose the interior of the sample, or to prepare a fresh surface, try a simple appropriate method: If the sample is soft enough, cut it with a knife or razor blade. If rubbery, use a pair of scissors or shears. If hard, try breaking, sawing, or abrading it. If the resulting surface is too rough or glary and refining the surface is inappropriate, wet it with an inert liquid and cover with a large-enough cover glass.

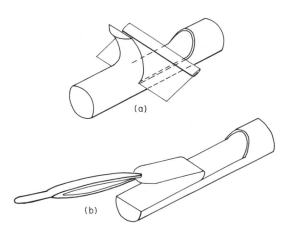

FIGURE 3.8. Peeling a fiber to yield (a) skin and (b) longitudinal section.[5]

   A great deal can be done under the stereomicroscope with un-
sophisticated tools. For example, R. B. Scott[5] simply taped an acrylic
fiber onto a microscopical slide under a stereomicroscope at about 60×.
Then with a hand-held razor blade he made a cut, as illustrated in
Figure 3.8a, and peeled off the skin which could be examined separately. By making a second cut (Figure 3.8b), he peeled off a longitudinal
section only 5–10 μm thick and examined it between crossed polars (see
Chapter 5). The success in making these longitudinal sections may be
attributed chiefly to the layered structure of the fiber that allows it to be
peeled, thin layer after thin layer. The phenomenon is all the more
remarkable considering that a filament of only about 20 μm in diameter
was operated upon under a limiting resolving power of about 5 μm
(Table 2.2). The attributes contributing to visibility are often surprising.[6]

3.5.   THE BIOLOGICAL MICROSCOPE

   In Chapter 1 we saw that historically biology was the first of the
sciences to grow up with microscopy, particularly through the
nineteenth century. This dual growth was quite separate from
developments in petrography and chemistry, where *different* requirements for the microscope were being met. Thus the distinctive instrument called the biological microscope appeared. It is a compound microscope, highly developed in its best models, in accordance with most
of the theoretical principles expressed in Chapter 2.
   In the brief period since 1950 the biological microscope has undergone a further period of intense development. In essence it has become
increasingly like the petrographical and chemical microscopes, with innovations that allow the use of polarized light and phase-amplitude
contrast. The resulting instrument is an all-purpose microscope which
reflects the blooming of science in general and microscopy in particular.
In this chapter, however, we are concerned chiefly with those attributes
contributing to visibility which make the traditional biological microscope so useful to the life sciences.

3.5.1.   Attitudes

   The *attitudes* in biological microscopy are understandably not only
those of the scientist, but also those of the technologist and the techni-
cian, chiefly in medicine. From premedical courses in biology, botany,
zoology, and biochemistry, to specific microscopical courses in medical
colleges and nursing schools, the microscope is thought of principally as
a *tool* for teaching and learning. Oftentimes the instructor sets up the

microscope with prepared specimens, supplemented by pictures of what his students are expected to see. Even in hospitals, doctors' offices, pharmaceutical houses, and medical institutions there is standardization of microscopes, illuminating devices, and preparatory techniques handed down from one generation to another with traditional attitudes. Only recently has there been a welcome change to the attitude that "microscopy needs better microscopists."[7]

3.5.2. Experience

Long *experience* has led to complex techniques for preparing biological specimens by dehydrating, "fixing" with alcohols, hydrocarbons, waxes or resins, staining, microtoming, mounting in more resin, and permanently sandwiching the product between glass plates. Some technicians become so expert that they prepare slides professionally, for use in schools. As a result, students become used to stereotyped appearances of mummified materials, and there is great temptation to copy other people's illustrations year after year, decade after decade.

3.5.3. Learning and Teaching

Only by *learning* more about microscopes and microscopy can biologists break out of this established pattern.[7] Only by *teaching* microscopy as a science in itself can biologists have prerequisites to laboratory courses in biology and medicine.[8-11]

3.5.4. Resolving Power

Biologists have a special need for the utmost *resolving power* in their microscopical objectives. Their needs stem from the small sizes of protozoa, bacteria, and structural cells (and their parts) in human, animal, and plant tissues. We have already seen that resolving power in objectives is bought in the market as *numerical aperture*,

$$NA = n \sin \alpha$$

where α is half the angular aperture in the medium of least-refractive index n. For dry objectives the least-refracting medium is air, so $n = 1$ and NA must be less than unity. Hence the biological microscope uses "high-dry" objectives of NA = 0.5 to 0.85 and enough working distance to examine tissues, starches, pollens, and larger structures. For higher resolution, immersion objectives (NA = 1.2 to 1.4) are especially useful for examination of such small units as bacteria and single cells. The traditional type is the oil-immersion objective. The "oil" nowadays is a nondrying, refractively stable, viscous liquid of refractive index 1.515 to match that of glass slides.[12] There is no use having an immersion liquid of n higher than 1.523 because that is the prescribed index for cover glasses.[12] However, for uncovered specimens in water there may be good use for a water-immersion objective (NA = 1.25) or a glycerine-immersion objective (NA = 1.25).[2]

3.5.5. Resolution

Resolution is obtained from the purchased resolving power (NA) *only if* the objective is used as intended by the manufacturer. First of all, he expects that dry objectives will not be immersed and that immersion objectives will not be used in air or other incorrect fluid. Second, the manufacturer expects a reasonably proper thickness of cover glass to be used with high-dry objectives. Number $1\frac{1}{2}$ cover glasses are specified by ASTM to have thicknesses of 0.17–0.19 mm.[12] The variation is permissible in the interest of economy. For "critical" microscopy the cover glasses should be 0.180 ± 0.002 mm thick unless the manufacturer of the objective specifies otherwise.[12] Another solution to the problem of variation in thickness among commercial cover glasses is to have an objective equipped with a correction collar.[2]

Even then, full resolution will not be obtained from a particular objective unless it is used with the proper eyepiece, condenser, tube length, and other appurtenances contributing to visibility.[13] Of these, emphasis is placed here on the proper condenser for the objective. To obtain full resolution, an oil-immersion objective requires an oil-immersion condenser and *the oil must be used.* Any condenser marked NA = 1.0 or higher is necessarily designed for oil immersion. It has the indicated NA only if the top lens is connected with the specimen by immersion oil. For instance, a condenser marked NA = 1.30 when used dry has an NA of only 0.86. It would be as well to use an air condenser marked NA = 0.90; in fact, it would be better corrected for aberrations than the oil-immersion condenser misused in air.[1]

3.5.6. Contrast

Contrast is often a special problem in biological microscopy because the differences in refractive indices between constituent parts of the specimen often are small, and are literally fixed[10] by traditional methods of preparing the specimens. In the past, contrast in postmortem specimens has been achieved acceptably by preferential staining or other physical–chemical treatment.[10] Such methods have become so customary that their effect on the original tissues is hardly realized. As in all other areas of microscopy, anything done to the sample alters it, and the drying and staining of samples is no exception. Some of those treasured structural features memorized by students may be just reaction products of staining. Some attention has already been paid to the effect on contrast achieved by the substage condenser and its diaphragm, together with the many other aspects of illumination[13,14] (Figure 2.11[6]).

Phase-amplitude contrast is especially appropriate for biological specimens in their aqueous media: bacteria, bacteriophage, blood cells, living diatoms, yeasts, and fresh (undried) cotton fibers are best observed this way. Phase-amplitude contrast[15] and dispersion staining[16] will be discussed in Chapter 9. The other methods of optical staining will be discussed in Section 3.3.11.

3.5.7. Corrections for Aberrations

Correction of objectives for *refractive* and *chromatic aberrations* is of specific importance in biological microscopy insofar as sharpness of image, wide field, color rendition, and other attributes are of special significance. Image sharpness[14] near the limit of resolution[17] in light microscopy perhaps is not as important now as it was before the advent of electron microscopy (1940). However, correct color rendition will continue to be important as long as electron microscopes continue to operate monochromatically, i.e., at fixed voltage and without visual color (Chapters 12 and 13).

Fidelity in the rendition of color images will also continue to be important as long as differential color staining is part of biology. This is quite different from the many significant colors in nature brought about by interference ("structural color") and by functional pigments, such as the colors in leaves, blood, skin, feathers, butterflies, and flowers.[2] Another reason why color rendition is so important in biology is that color photography and cinematography have been so highly developed that color in biological photomicrography is now a necessity.

In all cases where color fidelity is important, the objective and condenser should be corrected for chromatic aberrations. An aplanatic–achromatic condenser should be used with apochromatic objectives, or they cannot do their job properly. With achromatic objectives, ordinary uncorrected condensers with only two lenses or with a third (auxiliary) lens are appropriate. With objectives of lowest power, the auxiliary (upper-most) condenser lens should be removed; otherwise the iris diaphragm will not function properly.[1]

3.5.8. Cleanliness

Cleanliness in biological microscopy has more significance than mere removal and prevention of "dirt"; it also means exclusion of unwanted organisms. Since the days (Chapter 1) when contamination with microorganisms was interpreted as "spontaneous generation," unwary microscopists have been surprised by the appearance of organisms in contaminated specimens, reagents, instruments, and vessels. In addition, microscopists who use immersion oils must constantly be careful to keep objectives, condensers, and their hardware clean. At the same time, care in cleaning must be taken to avoid scratching glass, loosening lens cement, corroding metal mounts and mechanisms, and leaving hardenable liquids in the wrong places.

3.5.9. Depth of Focus

Depth of focus, as shown in Figure 2.9 and Table 2.2, decreases as the numerical aperture becomes larger. Therefore small depth of focus is a serious limitation in biological microscopy, since so many things of interest are near the limit of resolution. One small aid in this dilemma is an iris diaphragm placed in a small threaded cylinder which substitutes for the permanently open base of a separable objective of high NA.[2] With such a diaphragm the NA can be gradually reduced, oftentimes compromising successfully between resolving power and depth of focus.

3.5.10. Focusing

In *focusing* on the image produced by a biological microscope, the attitude should be one of *relaxation,* imagining that the object is far away at "infinity." Then, muscles of the eye are relaxed and tireless (see Figure 2.10). There need be no squinting over a monocular microscope, even by the unused eye. If that eye continues to sense a *macro*scopical image that cannot be ignored, cover the unused eye until you feel comfortable using only one eye. If your eyes still feel tired when using the microscope, stop a moment to look at a distant object so as to focus again at infinity. Do not attempt to bring a microscopical image into focus by accommodating with the eye's focusing muscles; they will get tired and you may get a headache.

3.5.11. Illumination

Illumination of the sample being observed in a light-transmitting microscope is provided by the *condenser.* In the biological microscope the condenser takes on special importance if the objectives are of especially high numerical aperture and are especially highly corrected (apochromatic). The condenser should match the objectives in both respects, or the objectives will not perform up to expectation and price. The condenser should be precisely centered at the factory or it should be centerable by the user. If it is centerable, it can be purposely decentered at will to obtain oblique bright-field illumination (see Figure 2.11). Surprisingly enough, a fixed condenser is not always well-centered at the factory.[11]

The condenser should also be focusable, for otherwise neither critical nor Köhler's illumination may be selected, and something unsatisfactory may be obtained as a result. It follows that the focusable condenser must be focused intelligently by the user, or it cannot fulfill its function.

The condenser in a biological microscope usually has a rack attached below the lenses to hold filters or dark-field (central) stops. Such a rack can also hold Rheinberg disks, like those shown in Figure 3.9, to provide one color for rays of low angular aperture and a different color for rays of higher angular aperture. The simplest type of differential color illumination is provided by a central disk of one color (usually a darker one for background) and a different (lighter) color for oblique rays which are scattered into the objective by the specimen. In Figure 3.9a, the central disk represents a blue (BL) filter the same size as an opaque disk that would give a dark field for the particular objective in use. The four annular quadrants are represented by Y for yellow. Figure 3.9b illustrates that the background would appear blue (BL) and a particle of specimen would appear yellow. In Figure 3.9c the central disk is opaque to provide a black (B) background. Two opposite quadrants are red (R) and the other two are green (G). Figure 3.9d represents a fabric of simple weave, oriented with respect to Figure 3.9c so that the warp is colored red (R) and the woof is colored green (G). More elaborate sets of disks have also appeared and then disappeared from the market. There is a similar method of optical staining in conjunction with the condenser which employs a small prism for the central rays and the microscopical mirror for the annular rays.[4]

For objectives of high NA, special dark-field condensers use reflecting surfaces to achieve angular apertures greater than those of the objectives, so as to obtain dark fields. Figure 3.10a shows a dry dark-field reflecting condenser and Figure 3.10b shows an oil-immersion dark-field reflecting condenser, for use with their respective objectives.[4]

3.5.12. Radiation

The most important *ultraradiation* in biology is that which excites fluorescence in the visible range.[18] For such radiation there are special

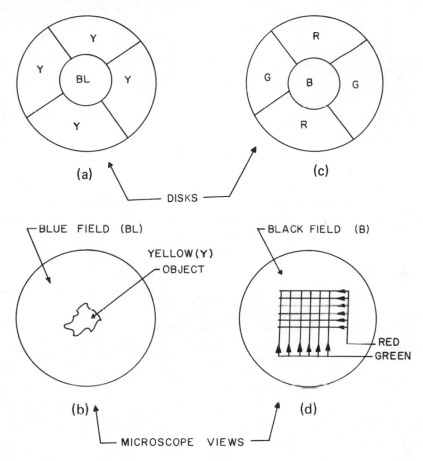

FIGURE 3.9. Rheinberg disks and their useful patterns.[4] Courtesy of Microscope Publications, Ltd.

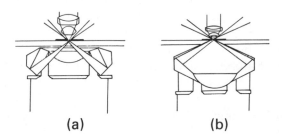

FIGURE 3.10. Light path through (a) dry dark-field and (b) oil-immersion dark-field reflecting condensers.[4] Courtesy of Microscope Publications, Ltd.

FIGURE 3.11. Reflecting objective.[1] Courtesy of Microscope Publications, Ltd.

objectives through which the rays are guided by *reflection*, rather than refraction. Like the condensers described in Section 3.3.11, reflecting objectives[19] are without chromatic aberrations throughout the entire spectrum of light, including not only the ultraviolet radiation but also the infrared,[20] as illustrated in Figure 3.11.[1]

3.5.13. Anisotropy

Anisotropy may exist in biological specimens in any or all of the manifestations described in Section 2.2.11. However, equipment for visibility of these phenomena is not usually included in a strictly biological microscope. The usual instrument has no polars, no rotatable stage, nor any slot for retardation plates; addition of these to a biological microscope is usually unsatisfactory, even if possible. Besides, the objectives and other optical parts provided on biological microscopes are not necessarily strain-free and will interfere with polarization phenomena. Therefore further discussion about anisotropy is reserved for Chapter 5 on polarizing microscopes.

3.5.14. Magnification

Magnification is still emphasized by biologists, rather than resolving power and the other attributes contributing to visibility. This is evident whenever biologists persist in using magnification as the first criterion in describing any objective, e.g., 40×, regardless of whether the objective has an NA of 0.65, 0.75, or 0.95, or whether it is achromatic, apochromatic, or in between.[1] Magnification has its place, especially in micrometry, but a distinction must be made between useful and empty magnification.[2] Magnification by itself does nothing to improve visibility.

3.5.15. Field of View

Field of view is important in biological microscopy because natural material is so complex and variable. Biologists know this, and they know how to cope statistically with the variables. Their experience with computers and the design of experiments makes them potentially good microscopists, who need only to understand more about the construction and use of their instruments.

3.5.16. Antiglare

Antiglare coatings are especially important on the air–glass interfaces of highly corrected objectives used in biology because there are so many such surfaces (see Figure 3.12[1]). Without such coatings the scattered light interferes seriously with contrast.

3.5.17. Cues to Depth

Cues to depth are especially important in biological microscopy. Biological samples are three-dimensional, but after thin sectioning the specimens are chiefly two-dimensional. Therefore it has been customary to make serial sections connected in sequence on a ribbon to be examined sequentially.[10] Reconstruction of the original structure in three dimensions from these serial sections puts an extra burden on the interpretive aspect of microscopy. Correlation of the result with information available from scanning electron microscopy certainly helps, if the instrument is available.

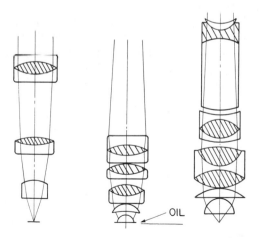

FIGURE 3.12. Typical microscopical objectives.[1] (a) 10× objective, (b) 61× objective, (c) 100× objective. Courtesy of Microscope Publications, Ltd.

3.5.18. Depth of Field

Depth of field may be a problem in biological microtomy when the specimen has to be cut thick enough to hold together and yet thin enough to get all of it into focus, top and bottom, with an objective of high NA. Table 2.2 shows that with a high-dry (NA = 0.95) or an oil-immersion objective, the depth of field is less than 0.1 μm, so the section has to be cut extremely thin. However, there are electron-microscopical techniques for making much thinner sections, and these techniques can be brought into light microscopy. The question that remains is what the technique may do to the specimen.

3.5.19. Working Distance

Special *working distance* has to be provided in biological microscopy, especially for thick cover glasses used on deep chambers in microscopical slides. Such chambers are used for counting blood corpuscles, growing molds, and examining embryos.[12] However, biologists have different microscopes available for these different purposes. For example, stereomicroscopes have long been used for dissection and are still called "dissecting microscopes," even though the newer biological microscopes can be used for high-powered dissection with mechanical manipulators.[4] Similarly, there are special microscopes designed for routine counting of blood samples and for pathological examination.

3.5.20. Structure of the Specimen

Many biological *structures* influence the formation of their microscopical images. Mention has already been made (Section 2.2.18) of the hexagonal pattern of the pores of *Pleurosigma angulatum* and its influence on the diffraction image of the skeleton (Figure 2.17).[13] Similarly, double-spiral configuration of fibrils in cotton has its influence on the images of cotton fibers examined under various conditions (Figure 2.15). The "nodes" in linen fibers of flax are due to the dislocations of fibrils in their fibrous bundles during thrashing of the flax, and are multiplied by wear and tear on the linen in use.[2] The characteristic appearance of muscle fibers and nerve fibers are related directly to their structures. Similarly, the "Maltese cross," common to all species of raw starch grains viewed between crossed polars, is related to the crystalline growth of the grains.

3.5.21. Morphology of the Specimen

The *morphology* (relative size and shape) of starch grains varies distinctively with the species, whereas the structure is common to all species. The microscopical difference between coarse and fine linen varies directly with the degree to which the fibrils have been separated from the original stalks of flax, and hence is a matter of morphology. The sizes and shapes of the cells of all kinds of natural materials may be observed under the biological microscope using the description for identification of fibers in textiles, paper products, and cordage.

3.5.22. Information about the Specimen

Of course, *information* about the specimens helps enormously in the identification of any material, but especially in products of nature which vary within species, among species, and within different commercial versions of various species. The information is all the more needed if the microscopist is limited to the use of only a biological microscope. The kind and degree of information required will vary, depending upon whether the microscopical examination is incidental or repetitive, scientific or technical, business or forensic, extensive or cursory, but it helps to obtain all the information possible.

3.5.23. Experimentation

Experimentation in biological microscopy may be limited not only by the biological microscope, but also by the accessories for it. Thus it does little good to equip the biological microscope with polarizer and analyzer to study the anisotropy of samples if the microscope does not have a rotating stage, and the effect of coating or abrading a surface cannot be observed if the microscope is not equipped for study by reflected light. Obviously, experimentation is facilitated greatly by using a universal microscope instead of the biological type.

Whereas "universal" microscopes are usually larger than the nominal biological one, the miniature microscope invented by a physician[21] can also qualify as a universal microscope. It can be equipped with polars and a rotating stage. The microscope is made the size and shape of a miniature camera by housing the tube in sections by means of prisms which also provide an erect image. The condenser stage and objectives are inverted with respect to the eyepiece. Since the thickness of the specimen slide does not need to be taken into account in focusing, the condenser and objectives may be set in focus. Any adjustments such as in optical sectioning is done by the fine adjustment knob; there is no coarse adjustment. Three objectives are mounted on a slide carrier for immediate use. It is a disadvantage that the objectives do not carry the ASTM standard threads, but fortunately the microscope has its own standard joints and fittings for quickly converting it from a portable field instrument to a more versatile laboratory instrument.

3.5.24.   Preparation of the Specimen

Any necessary *preparation* of biological specimens should be completely purposeful, employing modern methods[22–24] designed to minimize changes in the specimen. One of these methods may simply be freeze-drying of the sample instead of the complicated classical method of dehydrating, infiltrating, and fixing the sample.[10] Another method may be freeze-fracture replication[25] of the sample, which distorts it very little mechanically yet allows detailed study of the internal structure and morphology. Many other techniques adaptable to biological microscopy will be found in other chapters.

3.5.25.   Behavior of the Specimen

The word *behavior* has different meaning when applied to a biological specimen because the biologist is usually not concerned with properties like mechanical strength or modulus of elasticity. Instead, he wants to know about structure and development of an organism as functions of the life processes, and he prefers to study the living tissue under the microscope in its usual environment, if possible. Any device or accessory that will enable the biological microscopist to accomplish this aim is beneficial: a special chamber for studying perfused living tissue, a device for circulating nutrient liquid at controlled temperature, an arrangement for enhancing the visibility of a cell's nucleus against its surrounding protoplasm,[26] even a simple rimmed slide with its perhaps-too-thick cover glass. The aim is to study the behavior of the organism under different living conditions, including exposure to various agents. If the living organism itself cannot be studied on the microscopical stage, then its behavior has to be studied elsewhere and samples or sections have to be taken to the microscope. This introduces several more steps, so the behavior of the specimen under various methods of preparation must be investigated. All such experiments add information about the specimen as well as the method.

3.5.26.   Photomicrography

Photomicrography is just as important as photomacrography in biology, for just about the same purposes: records of growth and death of plants and animals, effects of environment on their tissues, time-lapse studies, high-speed cinematography, records by ultraviolet or infrared light, and special advantages of the photographic over the visual process, such as increased contrast. Every biological microscope intended for serious study should be equipped with camera or adapter, and with adequate illumination. The microscopist must also be familiar with the photographic materials available, and experienced in the use of them.

3.6.   SUMMARY

Visual acuity, which is the inherent resolving power of the human eye, increases (up to a point) as the eye is brought closer to the object. The limiting point is reached at a distance of about 25 cm, where the eye's lens is contracted as much as possible without strain. Since the limit of resolution of the eye is inversely proportional to its distance from the object, the only way to increase resolution is to supplement the lens of the eye with an external lens or lenses, thereby bringing the eye closer to the object than the natural limit of 25 cm.

The simplest device to accomplish this purpose is the common magnifying glass, which has its limitations: It cannot be held steady, the focus point keeps changing and is seldom optimum, and the image at the circumference of the lens is greatly distorted. When the lens is corrected and supported and brought into proper alignment and focus, it is called a *simple microscope* and is a very useful instrument. With it the limit of resolution is improved from 150 μm (for the unaided eye) to 10 μm (for the simple microscope of highest practicable NA).

The *compound microscope* allows increased NA and great improvement in resolution by using two or more lens systems. An objective (usually a multiple-lens system of short focal length) forms a tiny real (but inverted) image in the focal plane of a second lens system called the eyepiece, which magnifies that image to a point where it can be resolved on the retina of the eye. For optimum utility the lenses are corrected according to the optical principles of Chapter 2 and are mounted in a sturdy stand equipped with a light source, a condenser, a suitable stage for the specimen, a system of focusing, and a variety of accessories for micrometry, special illumination, micrography, and so on.

Compound microscopes vary in kind and extent of their contributions toward optimum visibility. The two general types considered in this chapter were the stereomicroscope and the biological microscope. Stereomicroscopes are very popular because they allow both eyes to be used for the perception of depth and are usually equipped to give an upright and hence easily interpreted image. Two designs are available, the first with two objectives, two eyepieces, and two body tubes inclined at a slight angle to each other so that the lines of sight converge on the object. This is the classical Greenough type, constructed with dry objectives of long working distance so that almost any object can be handled, turned, treated, and even dissected during observation. Since the left and right images are formed separately in axial lens systems, the systems can be fully corrected. One disadvantage is that both axes are inclined to the plane of the specimen, so that at maximum resolution only a portion of either field can be in focus at one time. A greater disadvantage is that there is no substage condenser, for the simple reason that it is impossible to shape a biaxial lens. Hence there can be no critical illumination, and resolution is severely limited. The other design of stereomicroscope can indeed have a substage condenser because it uses only one large objective, and splits the light from that objective into two parts which (after passing through image-erecting prisms) are observed through separate eyepieces in separate parallel body tubes. Both images are in simultaneous focus, and various accessories for different kinds of illumination are easily fitted.

The biological microscope is traditionally a compound microscope with a fixed stage, designed for examining transparent or translucent biological specimens and demonstrating their parts and structures. These instruments are capable of doing far more than is generally asked of them; students and researchers alike will benefit greatly from a study of the optical principles and the attributes contributing to visibility which are involved. Biologists need (and can obtain) extreme resolving power, attained by the proper matching of condenser and objective of high NA used with an immersion liquid of $n = 1.515$. If lower resolving power will suffice, it is better obtained from an objective of appropriate NA designed for air immersion than from an oil-immersion objective of higher NA used dry. Correction for chromatic aberration is important in obtaining true color rendition. Conditions of illuminating the sample and focusing both condenser and objective are important in order to obtain adequate contrast and sharpness without undue eyestrain. Dark-field illumination, oblique illumination, optical staining, and phase contrast can contribute much to visibility. Magnification is always secondary to resolution, and so the selection of objectives and eyepieces should be made with this in mind. Depth of field and working distance are also linked to numerical aperture, and an understanding of these relations can help the biologist to get much more from his microscope and learn more about the structure, morphology, and behavior of his materials.

Compound Microscopes Using Reflected Light

4.1. STUDY OF SURFACES BY REFLECTED LIGHT

Microscopy by reflected light may be used for a number of reasons: to look at a natural surface like that of a leaf, feather, skin, shell, or fossil; to compare surfaces after aging, usage, weathering, or other exposure; or to prepare an inside surface for studying an opaque substance such as bone, metal, coal, ore mineral, ceramics material, or pigmented plastic. A related thought is: How many layers are there in a seashell, tree's growth, laminated paper, or board? What is the structure of a sponge, tree cone, botanical cane, zoological organ, fossil, rock, ore (Figure 4.1), brick, cement, or plastic filled with biological material (Figures 4.2–4.5)?[1-5] Indeed, the specimen may be a particulate material such as seeds, tiny insects, sand, rock dust (Figure 4.6[1]), or small crystals. Such specimens may be better embedded in a dark, pigmented resin for reflected light than in a clear, colorless resin for transmitted light.

Experience, habit, or equipment may determine whether the light should be reflected from the prepared surface of a thick specimen or whether it should be transmitted through a prepared thin section. Metallographers,[2,3] ore-dressing engineers,[4,5] and resinographers[6-8] are experienced in the use of reflected light, while biologists are more experienced in the use of transmitted light.[9] In case the biologist has no choice but to use reflected light, he should gain sufficient experience to solve his microscopical problems.

The microscopist can *imagine* that reflected illumination may reveal more or different information than transmitted light. For example, anthracite coal may be petrographically prepared thin enough to be examined classically by transmitted light. However, a thick section may be

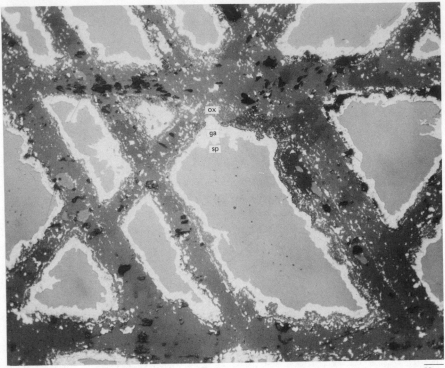

50 μm

FIGURE 4.1. Lead–zinc deposit, South Africa. Sphalerite (sp) replaced along its cleavage planes by lead–zinc oxidation products (ox) and by galena (ga). Polished thick section, reflected light. Courtesy of American Cyanamid Co.

examined by reflected light as polished or as etched, with or without polarizing the light. Figure 4.7 shows the differences under reflected light by varying the methods of examination.

With the increasing development of man-made composites of opaque and transparent, hard and soft, large and small components, organic and inorganic plastics, came greater use of reflected light on strictly transparent specimens[10,11] (Figure 4.8[12]).

4.2. RESOLVING POWER

The *resolving power* of microscopical objectives for use by reflected light can be slightly higher than those intended for use with transmitted

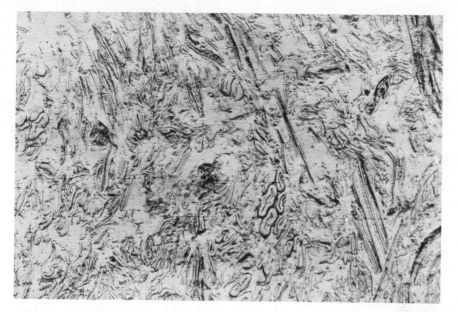

50 μm

FIGURE 4.2. Wood flour mounted in black phenolic resin. Polished cross section, reflected light. The fibers are straight in lengthwise section, and some cross sections occur in groups.[1]

50 μm

FIGURE 4.3. Sugar cane (bagasse) mounted in black phenolic resin. Polished cross section, reflected light.[1]

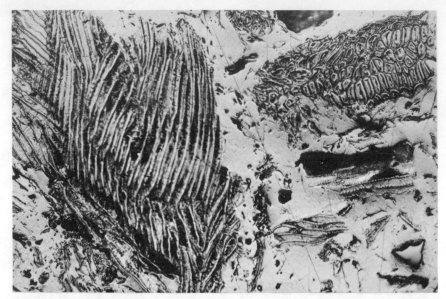

50 μm

FIGURE 4.4. Peanut shells mounted in black phenolic resin. Polished cross section, reflected light, showing plates of short, interlocked fibers in longitudinal and transverse sections.[1]

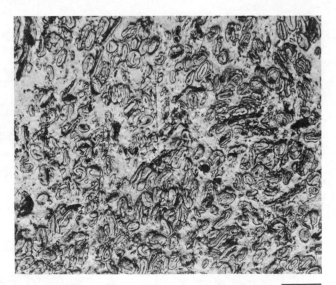

50 μm

FIGURE 4.5. Cotton mounted in black phenolic resin. Polished cross section, illumination by reflected light. Most of the fibers are shown in cross section. Oval cross sections are shown with slit in center. Actually, they are cross sections of *collapsed* cylindrical tubes.[1]

50 μm

FIGURE 4.6. Dust from Vermont chrysotile asbestos filler. Illumination by vertical reflected light. Gray particles are chrysotile asbestos or massive serpentine. White particles are associated magnetite. Matrix is Melmac® melamine–formaldehyde resin.[1]

light *and* a cover glass. No cover is necessary on a prepared surface being examined by reflected light, so the refractive index n in the formula for numerical aperture

$$NA = n \sin \alpha$$

is not restricted to 1.52 for glass or any other cover. Accordingly, an immersion objective of $NA = 1.60$ was made for use with monobromonaphthalene ($n_D = 1.66$)[11] as the immersion liquid, with corresponding high resolution. Unfortunately, this kind of objective is no longer manufactured; evidently there was insufficient demand for it, presumably because prospective users did not understand its optical advantages.

The resolution of fine structures by reflected light begins with proper objectives and their proper use.[13] If the objectives are corrected for use without a cover glass, no cover should be used. Reflected-light objectives are usually in short mechanical mounts for optical reasons, and bear the manufacturer's code of information. It is essential to know

FIGURE 4.7. Four ways to view a field of polished anthracite coal by reflected light: (upper left) between crossed polars; (upper right) unpolarized bright field; (lower left) unpolarized dark field; (lower right) bright field after etching with chromic acid.

the code and use the information. Is the objective corrected for infinity? If so, it is probably not interchangeable with those of a microscope of any other make (or, perhaps, other models of the same make). If an objective is corrected for use with a certain eyepiece or projection piece, use it. If an objective is to be immersed, be sure that the immersion fluid has the proper refractive index.

4.3. CONTRAST

Contrast in images by reflected light is low because of many partial reflections from various surfaces as the beam goes down through the objective to the specimen and back through the objective, illuminating reflector, and eyepiece. With highly reflecting specimen surfaces like those of polished metals and ores, there is still plenty of contrast. However, poorly reflecting, highly scattering surfaces like those of paper, textiles, and wood cause much glare.[11] Even so, as shown in Figures 4.1–4.8, enough contrast together with sharp resolution can be obtained from transparent material by reflected light to provide valuable information to those who will work for it.[12]

4.4. CORRECTIONS FOR ABERRATIONS

Corrections for aberrations, in dry objectives of NA = 0.5 or more for use by reflected light, are designed with the assumption that no cover glass will be used.[13] On the other hand, corrections for immersion objectives carry the assumption that any cover should have the same refractive index as the immersion fluid. Ordinary "oil-immersion" objectives assume that the "oil" will have $n = 1.52$, so ordinary glass covers may be used with oil-immersion objectives on occasions when covering will help the visibility. For example, the surfaces of paper, textiles, and wood scatter much less light if covered with the proper

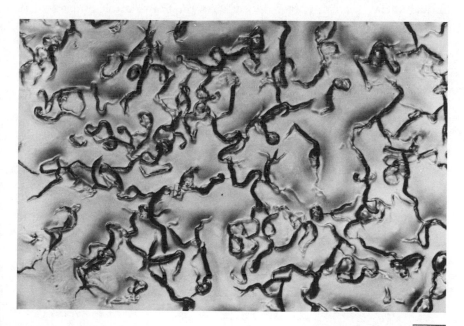

50 μm

FIGURE 4.8.   Wax on glass, cooled in about 20 sec. Micrograph by reflected light. Courtesy of Continental Oil Co.[12]

immersion fluid in conjunction with an immersion objective. If the specimen is fibrous, particulate, or dusty, the particles may be held in place by a proper cover slip and immersion liquid.

4.5. CLEANLINESS OF THE SPECIMEN

Cleanliness of the specimen is a special attribute contributing to microscopical visibility of a surface by reflected light.[5] Dust, abrasives, films, and even fingerprints will scatter light enough to add glare to the image and confusion to the interpretation. Precautions should also be taken with regard to the illuminating reflector inside the microscope, which is usually directly under the eyepiece or projecting unit. It is especially important that the tube *always* be covered to protect the illuminating surfaces from dust and dirt, and they must be inspected and cleaned periodically.

Cleanliness in a broader sense includes protection of the specimen from contamination and damage, care of the equipment, knowledge of optical principles, and order in general. All aspects of cleanliness are especially important in working with reflected light because so many optical surfaces are involved twice, first in illumination and again in image formation. Proper alignment of the parts and adjustment of diaphragm apertures are important for the same reason.

4.6. DEPTH OF FOCUS

Depth of focus is characteristically important in images formed by reflected light because the tilt of the illuminator is critical. If the tilt is not exactly 45° to the axis of the microscope, not all of the image will be aligned 90° to the axis. The eye has great depth of focus, and so if the depth of focus of the objective is great enough, all of the image may be in acceptable visible focus. Yet in photomicrography the eye is not directly involved, and so the depth of focus in the objective is critical in obtaining a good photomicrograph.

4.7. ROLE OF FOCUS

The *role of focus* itself is perhaps not as important in working by reflected light as by transmitted light, provided that the specimen's surface is flat and oriented exactly 90° to the axis of the microscope. Yet there are times when a deep etch or relief polish, plus unidirectional oblique illumination, is desirable in order to produce shadows. Then selection of the plane of focus is arbitrary, and therefore critical.

4.8. ILLUMINATION

Illumination is the most determinative attribute contributing to microscopical visibility by reflected light. As seen in Figure 2.11, reflected illumination can be of the dark-field or bright-field type; it can be unidirectional, symmetrical, or in between. Unidirectional dark-field illumination can be obtained from an ordinary reading lamp with a stereomicroscope (see Chapter 3), or with a smaller lamp and an ordinary compound microscope. Symmetrical dark field with such microscopes can be obtained with the Lieberkühn type of concentric, concave mirror (Figure 4.9[14]) or from a ring-shaped lamp fitting around the objective, such as the old-fashioned wreath of small incandescent lamps,[13] or from the newer circular fluorescent lamp, small enough to

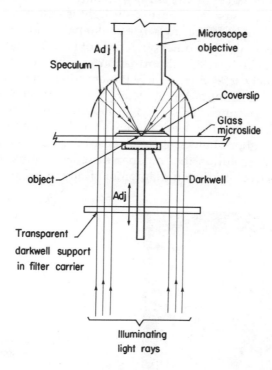

FIGURE 4.9. The Lieberkühn type of illuminator.[14] Courtesy of Microscope Publications, Ltd.

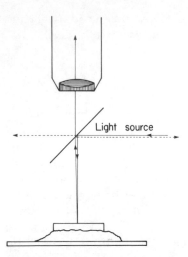

FIGURE 4.10. Vertical illumination by a transparent reflector between objective and specimen.[13]

fit close to the objective, or from a circle of fibrous light-wires[14] surrounding the objective. In any case, with dark-field illumination a mirrorlike surface appears dark because the light is reflected away from the objective, but a light-scattering surface like paper appears bright.

With stereoscopic and other objectives of long-enough working distance, bright-field (vertical) illumination can be obtained simply by placing a partially reflecting, clearly transmitting sheet (such as a thin, flat cover glass) in front of the objective at 45° to its axis and to the flat surface of the specimen, as shown in Figure 4.10.[13] The photograph Figure 4.11 was taken this way. It shows a ball bearing of high-carbon steel retained in both halves of its case-carburized steel race.

For higher resolution the transparent reflector must be placed beyond the objective. As shown diagrammatically in Figure 4.12, the cen-

FIGURE 4.11. High-carbon-steel ball bearing in its low-carbon, case-carburized race, mounted in bakelite, sectioned, polished, and etched. Illuminated as shown in Figure 4.10.

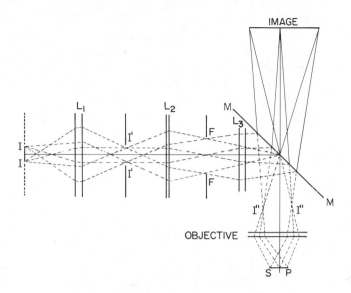

FIGURE 4.12. Vertical bright-field Köhler illumination by reflected light.[15] Courtesy of Microscope Publications, Ltd.

tral mirror MM is for bright-field illumination of the specimen SP, which returns some light back to the transparent reflector MM allowing light rays to pass through the system and form an image. Light from the source goes through a series of lenses and apertures and enters the side of the microscope. There the light strikes the transparent mirror MM and is reflected into the objective, which acts as a condenser. In the manner of Köhler's illumination, the field iris diaphram FF is focused in the plane of the object. The iris FF is opened just enough to fit the field of the particular objective. Light is reflected back from the object through the objective, this time through the mirror MM to form a real bright-field image.[15] This image is enlarged by the eyepiece or projection piece (neither shown). By shifting FF and/or tilting MM, oblique bright-field illumination is caused to form shadows on ups and downs in the object.[1,13]

Figure 4.2 illustrates the use of slightly oblique reflected illumination in observing a transparent object.

For comparable dark-field illumination by reflection from the specimen, a ring-shaped condenser is put around the objective, as indicated in Figure 4.13.[1] This figure also shows the annular mirror AM, which reflects light from the source to the annular condenser AC. Manufacturers have various arrangements for shifting between bright-field and dark-field illumination. The most convenient and quickest changes

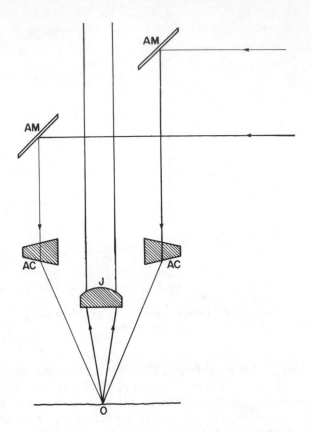

FIGURE 4.13. Dark-field illuminator by reflected light. A specularly reflecting object reflects light outside the objective so that the specular surface appears dark (hence dark field). A rough surface, however, scatters some light into the objective and appears bright.[1]

involve special objectives with special threads. The illuminators are special too.[11,15] The larger microscopes, including metallographs,[1] employ prisms, rather than glass plates, as reflectors. Such blocks of glass vary the tube length so as to upset the corrections in objectives for aberrations. Consequently there are objectives corrected for imaging at infinity. They use an auxiliary lens placed behind the thick prisms to bring the image into focus, as indicated in Figure 4.14.[15]

4.9. RADIATION

The *radiation* most used by reflecting microscopes continues to be that from an incandescent tungsten filament. Although a broad tungsten ribbon filament gives a more desirable and more uniform source of light, the coiled tungsten wire filament prevails on the market. The intensity of any incandescent filament is best varied by controlling the input voltage by means of a variable transformer (not by manipulation of aperture or field diaphragms!). Lowering the voltage does change the color temperature toward the red end of the spectrum. A bluish daylight filter will stabilize the color of the light and help in stabilizing colors in the image. With colorless images it is traditional in metallography to use a green filter with its transmission peak at or near the wavelength for which the objective was chromatically corrected.[16]

The Pointolite® source is a tungsten arc encased in a glass bulb containing an inert gas. It is of high intensity and of high actinic value in all regions of the spectrum, including the ultraviolet. A tungsten arc operating in mercury vapor is even stronger in the ultraviolet.[13] If only the ultraviolet region is desired, the visible portion is removed by an appropriate filter; likewise, if the ultraviolet is to be removed in order to look at the beam or source for any reason, an *adequate filter* or pair of goggles should be used *to protect the eyes.*

The carbon arc lamp is the traditional source of radiation in photomicrography by visible or ultraviolet reflected light. Modern carbon arc lamps have automatic carbon feeding and run either on direct or alternating current (although the dc arc is the more brilliant, if the much hotter positive crater is properly put into focus).[14] Commercial carbon rods are either for dc (thick for + and thin for −) or for ac (one size). The carbons come with a variety of cores composed of chemical compounds chosen to emit a narrow or broad spectrum, as required.

The newer metal–halide lamps, too, contain metal additives in proper proportions to give a white light with high color fidelity.[14] The quartz–iodine–tungsten lamp also has the high intensity needed in the reflected-light microscope, where so much light is lost by transmission, reflection, or light-scattering.[5,17]

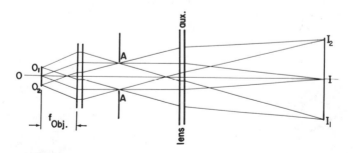

FIGURE 4.14. Image formation by an objective, corrected for "infinity."[15] Courtesy of Microscope Publications, Ltd.

Another lamp of high intensity is the xenon arc. Unlike the quartz–iodine tungsten (coiled filament) lamp, the xenon arc lamp gives off a solid spot of light.[17]

So far, the main reason for demanding a light source of high intensity for microscopy by reflected light is that so much of the light is lost before the image is finally formed. Another important reason is that the microscopist often wants to select only certain portions of the spectrum and, after allowing for unavoidable attenuation, must still have enough intensity left to form an image. One example is in fluorescence microscopy by conversion of an incident beam of ultraviolet radiation into visible light, as shown diagrammatically in Figure 4.15. Since fluorescence occurs at the surface of the specimen, it is an advantage to use incident ultraviolet light as shown, instead of a transmitted beam,[14] because the specimen need not be thin. In the case of powders, paper, textiles, or surface coatings, little or no preparation of the sample is necessary.

Many ore minerals and some metals show natural color, and many more may be colored experimentally. These and other specularly colored specimens may be compared directly on a comparison microscope,[13,14]

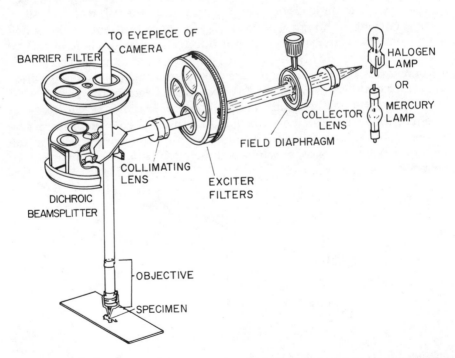

FIGURE 4.15. Fluorescence by conversion of ultraviolet to visible light. Courtesy of the American Optical Company.[14]

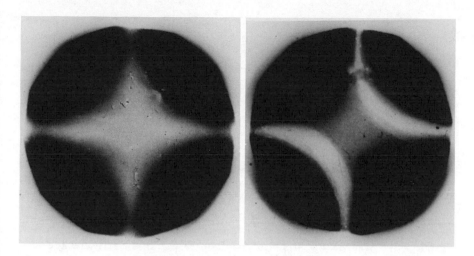

FIGURE 4.16. Figures obtained by reflected light with pyrite (isotropic, left) vs. arsenopyrite (anisotropic, right), with an objective of high NA. Courtesy of American Cyanamid Co.

with standard specimens or with color standards. Nelson has converted a comparison microscope into a visual tristimulus microcolorimeter for the microscopical identification of a polished ore mineral by means of its specular color.[5]

For accurate measurement of reflectance, nearly monochromatic light is required. For most routine work four interference filters, each with a bandwidth of less than 25 nm, are adequate. Their peak wavelengths are 470, 546, 589, and 650 nm. An interference filter of continuous wavelengths (400–700 nm) allows the construction of a dispersion curve (reflectance vs. wavelength).[5]

Many minerals and other crystalline materials are anisotropic. This means that reflectance and color both vary with the vibration direction of polarized light with respect to the orientation of each crystal grain. Anisotropic crystals also have positions of extinction between crossed polars (Chapter 5). These phenomena mean that polished anisotropic crystalline grains present various kinds and degrees of visibility as they are rotated between crossed polars. At the back aperture of an objective of high NA an interference figure may be seen[18] (Figure 4.16).

4.10. MAGNIFICATION

Exact *magnification* is determined by reflected light in the same way as by transmitted light[14]: by comparing the real or virtual image with

the graduations of a standard scale placed as the object on the microscopical stage. However, by reflected light the spaces between
graduations on the standard scale must be specularly reflecting (silvered) instead of transparent (clear). Real images of the standard scale,
such as on a ground-glass projection screen or photomicrograph, are
measured in enlargement by a commensurately larger scale ("ruler").
Virtual images (those observed directly in the eyepiece) are measured by
means of a secondary micrometer (reticle) in the eyepiece. Such a micrometer must be calibrated against the standard scale placed as the
object on the stage for the particular objective, tube length, and
eyepiece. During calibration be sure that the unit graduations at both
ends of the scale have the same value (within your tolerance) as those
graduations in the middle of the scale. If not, return the scale to the
manufacturer for one that does. For very accurate work an institution
should have at least one stage micrometer which has been checked and
corrected by the National Bureau of Standards, Washington, D. C.
20234.

Magnification is essential in all kinds of micrometry. Figure 4.17 is
typical of the measurement of the layers on a case-hardened low-carbon

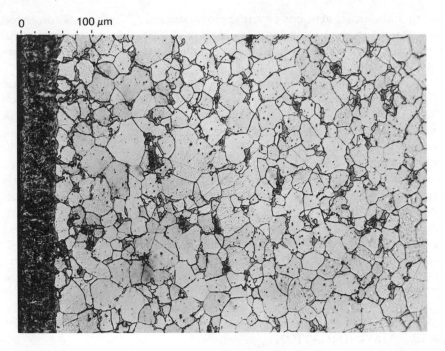

FIGURE 4.17. Bright-field, vertical illumination of a case-hardened specimen of very
low-carbon iron.

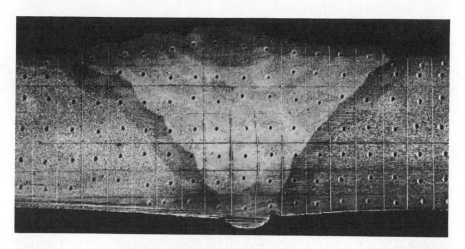

FIGURE 4.18. Hardness differences in a weld by means of indentation. Courtesy of A. White.

steel. The scale at the measured magnification is on a negative image of the specimen during printing.

Knowledge of the magnification is essential in the estimation of grain size in metals and alloys.[19] Likewise the magnification must be known accurately in measuring the microhardness of materials[20] (Figure 4.18).

4.11. FIELD OF VIEW

The field of view should be large enough to tell the story, as in Figures 4.17 and 4.18. On the other hand, a picture should be cropped of extraneous or distracting portions if they are on the periphery. By the same token, the most important part of a picture should be placed in the center of the field of view, if that is feasible.

4.12. PROBLEMS OF GLARE

Glare from reflected illumination is much greater than that from the transmitted variety. The reason is that there is much more attenuation of the light as it goes through the objective (or dark-field condenser) to the object and back through the objective. Even in the least complex system there is some scattered light. The problem gets worse as we add lenses, thick glass prisms for illumination, polars, interference filters, and so on. To see how much glare there is, focus on the lighted specimen, then remove it, and with the light still on, look again. You may see as much as one-third of the original intensity, due to scattered light. Removing the eyepiece and looking down the tube may reveal specific reflections from metallic holders or frames for the reflector, objective, polars, etc., suggesting that some jet-black paint in the right places may help in removing glare.[13] Perhaps the illumination is not centered or the field diaphragm has not been closed so as to frame the field of view and exclude stray light (Köhler illumination).[13,15]

The intensity of partially reflected light such as from the glass-to-air interfaces of objectives, field lenses, and filters can be reduced by coating such interfaces with a thin film of a substance (such as magnesium fluoride) which is less refractive than the glass. The intensity of partially reflected light at the illuminator may be increased by coating it with a substance more highly refractive or more highly reflective (e.g., silver).[11] Oftentimes, reducing the aperture back of the objective to cut out the most oblique, glare-producing rays will restore acceptable contrast. Of course, another objective of lower NA will do the same thing, but at lower objective magnification.[15] Scattering at the surface of the object can also be reduced by covering it with immersion fluid and using the proper immersion objective. On the other hand, visibility *may* be better by dark-field illumination, since the scattered light rays are then purposely collected by the objective and turned into the image, while the specularly reflected rays are rejected.[11]

Some of the reflected light is polarized, and therefore can be eliminated by orienting a polar so that its direction of vibration is crossed with the direction of vibration of most, if not all, of the polarized glare (see Chapter 5). The optimum effect is obtained by rotating the polar or specimen empirically until the glare is minimized.

4.13. CUE TO DEPTH

The best *cue to depth* by reflected light is attained by casting shadows inside depressions and outside elevations, shallow as they may be. This can be done with bright-field illumination by tilting the mirror. With dark-field illumination shadows are created either by shutting off half of the dark-field beam or by decentering the hollow mirror. Either way, relief polishing or deeply etching the specimen will lengthen the shadows.

4.14. DEPTH OF FIELD

Of course, etching or relief-polishing must not be so severe as to exceed the *depth of field* of the particular objective. This requirement is usually achieved when the specimen is prepared first as a smooth plane and then etched or relief-polished lightly and progressively until the desired effect is obtained. Usually a flat surface attained by commercial planing or rolling yields an image within the depth-of-field limitations of the optical system without further treatment. This is true of most samples of sheet metal, wooden boards, laminated paper, and cardboard, with or without paint, lacquer, varnish, or other finish. Of course, there will always be some specimens that are too rough for a particular objective's depth of focus. The task then is to prepare a smoother surface that will remain significant and representative.

Lack of depth of field with high-power objectives on upright microscopes which use reflected light can also be a problem if the prepared surface is not perpendicular to the axis of the microscope. If the base of the specimen is sufficiently flat, a partial shim under the low side or corner may be all that is needed to bring all of the field into focus. Mechanical clamps are available, or can be made, for holding the surface up against a hole that has its frame parallel to the base. For low powers, at least, an irregularly shaped specimen may be propped up with a little molding clay and a modicum of patience. With *inverted* microscopes and metallographs (Figure 4.19) the specimen rests on its plane-polished surface against the stage, which is perpendicular to the axis of the microscope.[11]

4.15. WORKING DISTANCE

Working distance between specimen and objective for reflected light is usually not a severe problem. One reason is that as a rule there is no cover glass (≈ 0.18 mm thick, or thicker). Another reason is that the specimen is usually smooth and flat without peaks to take up working distance.

Considerable working distance is needed in performing many kinds of operations and tests by reflected light, so provision has to be made. For example, in Figure 4.20 space is provided for a dental drill to remove, for microchemical test, a certain particle in a complex mixture of ore minerals and gangue in a polished section.[1] Incidentally, the micromanipulator is used to control other tools such as a tiny magnet, or dissecting needles, or syringes. Another kind of test which requires

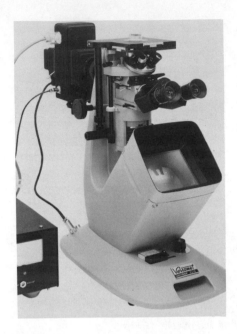

FIGURE 4.19. The microscope is inverted and illuminated with a tungsten lamp or a xenon arc. Courtesy of Unitron Instruments, Inc.

FIGURE 4.20. Dental drill and micromanipulator. The chuck of the dental drill is held in a special clamp attached to a micromanipulator.[1]

considerable working distance is that for microhardness,[20] either by scratching or indenting with a diamond. Measurement of the width of scratch or indentation also requires considerable depth of field.

4.16. STUDY OF STRUCTURE

The *study of structure* by reflected light is the very basis of metallography,[2,21] ore mineralogy,[4,5] resinography,[7] wood technology,[10] and a host of other technologies. In metals, alloys, ores, and some plastics we often have granular structures and sometimes fibers. In natural tissues such as wood, reeds, bark, and shells we have characteristic cells and fibers. The visible *kinds* of structure and their distribution are of fundamental importance. We must also consider the *orientation* of these structures, as shown in Figure 4.21.

Crystal structure is of fundamental importance to visibility. Practically all metals, alloys, minerals, and starches are crystalline. Consequently there are certain properties and phenomena that derive specifically from the crystalline structure of such materials. For example, in Figure 4.21 the granular structure of a metal is brought out by etching, principally because of the differential solubility of crystals in different orientations within the mass. Such characteristic orientation of the crystal grains is inherent in the crystalline structure. Likewise the direction of the slippage of the planes of atoms in a crystal of stressed metal varies with the crystallographic orientation of each grain with respect to the direction of stress.

Although the potential visibility of structure lies within the specimen, other attributes depend on the microscopy and microscopist. In Figure 4.21 visibility of the structure is aided by adequate resolution, contrast, shadows, field of view, preparation, and photography.

FIGURE 4.21. Stainless steel 18Ni + 8Cr, cold-rolled, polished and etched, showing one, two, or three slip directions, depending on the specific orientation of each crystalline grain with respect to the direction of cold-working.

4.17.   STUDY OF MORPHOLOGY

Visibility of the internal features of the metal in Figure 4.21 is aided by appropriate *morphology:* size and shape of the grains, distance between slip lines (representing the slip planes in section), and thickness of intergranular material. In a similar way, the morphology of the various phases in heat-treated, plain-carbon steels is made visible by reflected-light microscopy. This has enabled metallographers to understand greatly the principles of precipitation-hardening, as is summarized in Figure 4.22, which illustrates the metallography of heat-treating 0.89% carbon steel at various rates of increasing and decreasing temperatures. On the other hand, the precipitation hardness of aluminum alloys was not understood by light microscopy because the morphology of the precipitates was too fine. Nevertheless, experience with steels and brasses led metallographers to imagine what to look for using electron microscopy, and then their efforts were successful.

4.18.   INFORMATION ABOUT THE SPECIMEN

Information about any specimen is an attribute contributing to visibility. In microscopy by reflected light, there is hardly an exception. Usually sufficient information can be obtained firsthand to enable the microscopist to decide how he will go about solving his present problem. If there is insufficient information, it is best to go to the primary literature about the material before starting the investigation.

Figure 4.19 shows a microscope (among several varieties) which allows quick choice between reflected and transmitted illumination. Figure 4.23 shows a table model of a polarizing microscope which also offers a choice. Whether reflected or transmitted light (or both) are used depends on the descriptive information accompanying the sample, what microscopical experiments are to be performed, the behavior of the specimen on the microscope and in the room during the scheduled time, and how much preparation is necessary.

4.19.   EXPERIMENTATION

Microscopy involves *experimentation* as well as observation, and some possible experiments are suggested in Figures 4.7, 4.15, 4.18, and 4.20. Almost any experiment feasible by transmitted light may be performed by reflected light, with the added advantage of having the specimen uncovered so as to be able to pick, poke, scratch, cut, turn, or roll it. A small magnet, a tiny "soldering iron," a thermocouple, or any other active point may be applied. The specimen may be placed on a hot[13] or cold stage,[13] and it may be stretched, pushed, or twisted.

4.20. BEHAVIOR OF THE SPECIMEN

The *behavior of a specimen* during observation may be fortuitous, since it is uncovered and often exposed to a hot beam of intense light. Nevertheless, any change in the appearance should be noticed, recorded, and explained. The uncovered, freshly prepared specimen may have been stored a long time in an ordinary or extraordinary atmosphere, dry or humid, before or after etching, and it may have undergone change. Differences in behavior should be noted just as though the exposure were planned, as well it might have been.[10] Some of the reactions to look for are efflorescence, deliquescence, oxidation, sulfiding, and polymerization.

The chief limitation of microscopical examination by reflected light is the condition of the surface, so any change in that surface is significant. It must be flat enough and smooth enough for whatever objective is required to do the job. This is why a preliminary look at low power is so important. Even if the problem is not solved then and there, one can get an idea of how well the surface, as received, is suited to reflected illumination, how much preparation will be needed to see more, and how the surface has changed during storage and observation.

4.21. PREPARATION OF THE SPECIMEN

If the surface is dusty or dirty, clean it with a stream of air or some inert liquid. The safest liquid is water plus a little wetting agent, perhaps using a brush, soft at first, stiff if needed. If the surface has too many scratches or pits, and if it is hard enough, a fine abrasive (1 μm or less) may improve the image. But any substantial abrasion will reveal the interior rather than the exterior.

To expose the interior of a sufficiently solid, cohesive material, one may proceed to cutting, smoothing, and polishing. This is the way most metals, many rocks, ores, ceramics, cements, glasses, and plastics are prepared. If a specimen is powdery, particulate, fibrous, thin, porous, or irregularly shaped, it should be mounted by molding it in a block of resin. A molding press and mold are available for compression-molding bakelite and similar resins, providing a convenient method of mounting

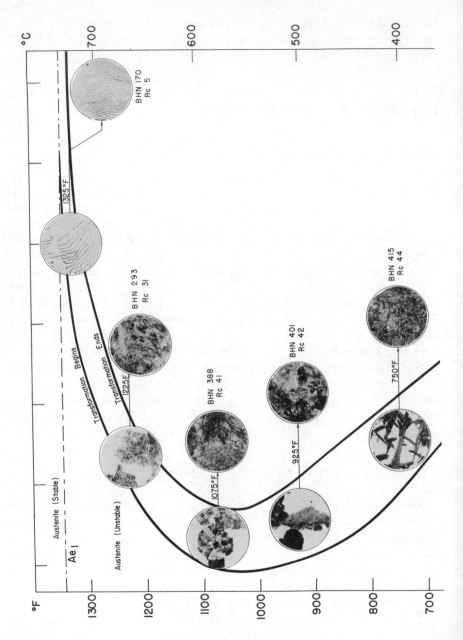

FIGURE 4.22. Isothermal transformation diagram of eutectoid carbon steel. C, 0.89%; Courtesy of U.S. Steel Research Laboratory.

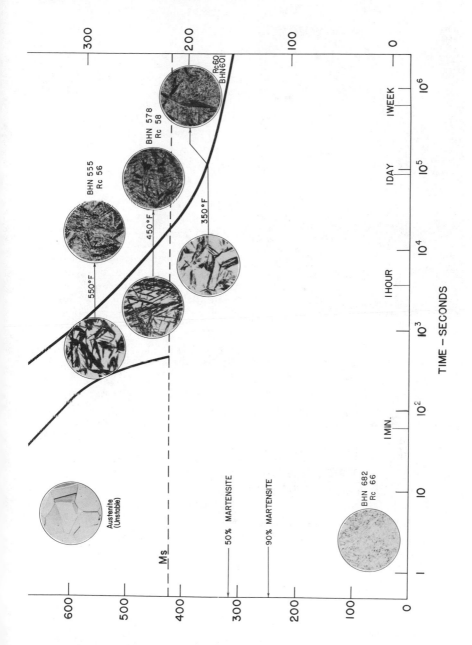

Mn, 0.29%. Austenitized at 1625°F. Grain size: 4–5. Photomicrographs originally 2500×.

FIGURE 4.23. D. W. Davis using an all-purpose polarizing compound microscope, used in industrial microscopy with either reflected or transmitted illumination, bright field or dark field. Courtesy of American Cyanamid Co.

the sample.[1] If compression molding is not feasible, casting in a polymerizable liquid is recommended.[22,23]

Etching is the key to the light-microscopical examination of all metals. Its primary purpose is to remove metal that has been modified by the preparatory procedures.[13,14,24]

4.22. PHOTOMICROGRAPHIC TECHNIQUES

The *photomicrographic techniques* by reflected light are largely those of general photomicrography and of metallography. A special problem may be poor contrast between those phases which do not differ much in reflectivity. Improved contrast can be obtained by using high-contrast photosensitive materials, and by using a high-contrast developer under extreme conditions.[16]

4.23. SUMMARY

Many materials are opaque, and therefore the microscopical examination of them must be accomplished by using reflected light rather than transmitted light. For example, a metal cannot be examined by transmitted light no matter how thin a section is prepared or how thin a foil is chosen; reflected light offers our only chance to learn something about the structure and morphology of the metal. In other situations, it may be only the *surface* of a transparent or translucent material that is important, and so examination of that surface by reflected light is the only way of going about the study. For all such cases, equipment and techniques for microscopical observation by reflected light have been developed to just as high degree as those for transmitted light.

The simplest optical arrangement for study of a specimen by reflected light is a polished glass plate mounted between the objective and the specimen at an angle of 45° to the axis of the microscope. Light from a source off to the side, at right angles to that axis, is reflected onto the sample. The different components of the specimen's surface then reflect the light in different and characteristic manners up through the glass plate and into the optical system of the microscope, forming an image of that surface. Obviously this arrangement will work only for objectives of low power and long working distance. For more detailed study the 45° reflector must be positioned in back of the objective, within the microscope body. The reflector may then be a mirror, a prism, or even a polarizing crystal positioned to reflect light downward on the sample. One advantage of this arrangement is that the objective automatically becomes condenser as well as objective (and a better-corrected condenser of higher numerical aperture than most external ones), so that excellent illumination is assured. Furthermore, in most instances a cover glass is not necessary, so that an immersion liquid of suitable refractive index may be used between the objective and the sample, thereby improving the resolution.

Light sources for reflective microscopy must be more intense than those used in transmissive work, because of greater losses in the optical system. Tungsten–halogen lamps, mercury or argon arcs, and even old-fashioned carbon arcs are preferred. The illumination may be bright field or dark field, according to the various optical arrangements described in the text and shown in the diagrams. With sufficiently intense sources and with appropriate filters, advantage may be taken of some special spectral portion of the illuminating radiation to produce differential fluorescence (or other exaggerated response) within the components of the sample.

One drawback of reflective microscopes is that part of the light is reflected or scattered by the many surfaces it meets on the way to and from the sample, so there is likely to be extraneous light (called glare) which reduces contrast and impairs visibility. The amount of glare can be reduced by scrupulous cleanliness of the optical components, by hunting down and blackening any reflective metal parts in the microscope, by using coated lenses, and (if absolutely necessary) by constricting apertures or installing diaphragms to cut out the wide-angle reflected light. Since such light is partially polarized, it can be reduced in intensity by installing an eyepiece polar and rotating it for maximum effect.

Not only existing surfaces, but also the *interior* structure and morphology of opaque materials such as bones and teeth, can be studied by reflective microscopy. Suitable sections are sawed or cut, and the fresh surface is smoothed and then polished until all scratches are invisible and the surface features can be seen. Etching the polished surface of a metal is essential in distinguishing the structural components and studying the morphology. Etching also helps in the study of plastics, polymers, minerals, and ceramics. Reference to a complete list of etchants for metals is given, and precautions for their use are included.[24]

5

Microscopy with Polarized Light

5.1. THE OVERHEAD PROJECTOR

The historic observation of double imaging by a highly birefringent crystal such as calcite (mentioned in Chapter 1) can easily be repeated on an overhead projector. A small dot of black paper is pasted on the projector's window near its center. Then a cleavage rhombohedron of clear calcite is placed over the dot. Two images of the dot appear on the projection screen, which means that the incident beam is being divided into two beams that do not interfere with one another because they are vibrating in different (perpendicular) planes. When the calcite rhombohedron is rotated on whatever face it happens to rest on, one image of the dot is stationary while the other image curiously traces an ellipse around the first image (Figure 5.1). The fact that the so-called extraordinary image is displaced from the ordinary image means that the extraordinary ray travels at a different velocity than the ordinary ray. The fact that the extraordinary ray traces an ellipse rather than a circle means that the *difference* in velocities varies with the orientation of this crystalline species. The ordinary image is stationary because its velocity is constant with all orientations of the crystal, and accordingly the ordinary refractive index ω is constant. The *various* values for the refractive index in the path of the extraordinary ray may be given the general symbol such as ϵ' or specific symbols such as ϵ_1, ϵ_2, ϵ_3, etc.

The overhead projector may be converted into a polariscope[1] by placing over the illuminated window two polarizing sheets (polars[1]), such as Polaroid®. The two polars are separated vertically by about 10 cm by means of two blocks so that the specimen may be placed and turned by hand between the polars. If the polars are crossed with respect to their directions for vibration of light, the field will be black

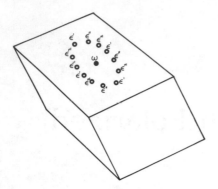

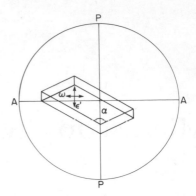

FIGURE 5.1. Single crystal placed over a single dot, showing double imagery: (1) The central imaged dot is stationary, since it is represented by the ordinary ray ($n = \omega$) inside calcite. (2) The elliptical ring of imaged dots indicates that there is an elliptical path as the calcite rhombohedron is rotated over the single object dot ($n = \epsilon_1, \epsilon_2, \epsilon_3$, etc.).

FIGURE 5.2. Calcite rhombohedron in one of four analogous positions of extinction (darkness): when vibration direction ω or ϵ' in calcite corresponds to vibration direction PP in polarizer or AA in analyzer.[2]

because none of the polarized light from the first polar (polarizer[1]) is vibrating in the direction for the light to vibrate through the crossed polar (analyzer[1]). An anisotropic crystal, such as a calcite rhombohedron, will appear bright in all positions of rotation except the four times in a revolution when the direction for ω or ϵ' corresponds to the polarizing direction of the polarizer or analyzer, as indicated in Figure 5.2. In all other positions of rotation the crystal appears bright because there is a vector corresponding to the direction of vibration in the analyzer, so that some light appears.

The extinction displayed by calcite, shown in Figure 5.2, is called symmetrical extinction because at extinction the obtuse angle α is bisected by PP or AA, the direction of vibration in the polarizer or analyzer. Some other species of crystal display parallel extinction which occurs when PP or AA corresponds to a prominent edge of the crystal. Otherwise an anisotropic crystal manifests oblique extinction. This and other optical properties in microscopic crystals call for microscopical examination by means of a polarizing microscope. It is employed in optical crystallography,[2-5] mineralogy, petrography, chemistry, physics, and biology.[6] Anisotropy is also observable in liquid crystals,[7] strained glasses,[8] stressed plastic materials and models, partially crystallized resins and polymers,[8] reflecting surfaces, flowing colloids, man-made filaments, and biological fibers, cells, and tissues.[9,10]

5.2. ANISOTROPY

Anisotropy may be classified into five kinds:

1.   The optical anisotropy[1] of each *single anisometric crystal* is manifestation that the spacing of the unit molecules, ions, radicals, or elements is different in at least two of the crystallographic directions. Crystals such as calcite, quartz, melamine, and sucrose display 10 or more individual optical properties that can be detected and measured on such microscopic crystals, properties which are very useful in characterizing materials in synthetical and analytical science and technology, as discussed in Chapter 7.

2.   *Multiples* of anisotropic crystals have optical characteristics above and beyond those of the individual crystals. The simplest multiples are *twins* (Figure 5.3), which are two crystals sharing a single "composition" plane,[2] but also manifesting individual properties. Figure 5.4 shows twins and higher multiples of subcrystals within crystals. Spherulites[1] are of still higher multiples of crystals, each one contributing to a unit effect on polarized light. Such spherulites are manifested, for example, by a single grain of raw starch or of polyoxyethylene.

3.   *Molecular birefringence* is manifested by long or flat molecules, especially by macromolecules (polymers). In *long* molecules the atomic dipoles are mostly arranged in chains. By induction the dipoles are stronger than if the atoms were widely separated. When polarized light is vibrating lengthwise to the chain, the average strength of the dipoles

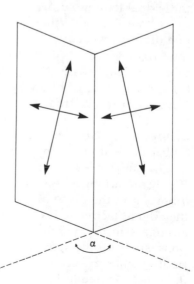

FIGURE   5.3.   Twinned   crystals   (with separate extinction directions).[2]

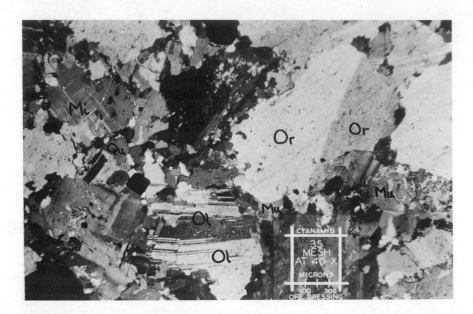

FIGURE 5.4. Thin section of pegmatitic rock (granite), between crossed polars, showing crystalline grains. The large grain Or–Or is a twin. The grains Ol–Ol and Mi are multiple crystals. Ol, oligoclase; Or, orthoclase; Mi, microcline; Mu, muscovite; Qu, quartz. Courtesy of American Cyanamid Co.

is greater than when light is vibrating crosswise. The optical properties of a system of parallel long molecules are derived from those of the individual long molecules.[7]

Primarily, light vibrating parallel to a molecular chain or to the axis of a (fibrous) system of aligned long molecules will travel more slowly than light vibrating crosswise. That is, the refractive index for light vibrating lengthwise ($n_\parallel$) is higher than for light vibrating crosswise ($n_\perp$). This is conventionally called positive birefringence in fibers (Chapter 6) and is strong in natural cellulose fibers and many man-made fibers. Of course, *side chains* on the molecule tend to reduce the strength of birefringence of the main chain of the molecule, and therefore that of a fiber made up of such branched molecules[7] (Chapter 6).

Flat molecules tend to slow down light that vibrates in their planes. If they are arranged with their planes parallel, as in a film or foil, light vibrating in the plane of the arrangement has a refractive index higher than the refractive index of light vibrating perpendicularly to the planar arrangement. Symmetrically planar molecules (such as ring configurations) are isotropic in their planes, so that films or foils made of symmetrically planar molecules are isotropic when viewed perpendicularly to the sheet. Although long *flat* molecules tend to be anisotropic in their planes, if they are oriented at random, the resultant sheet would still be isotropic in its plane. However, if the long flat molecules are oriented with their lengths parallel and their planes parallel, their sheets tend to be birefringent in their plane as well as perpendicular to their plane.[7]

If the long flat molecule also preferentially *absorbs* polarized light *strongly* in one direction (generally lengthwise), the film made with such molecules tends to be polarizing like Polaroid® polarizing film.

4.   *Form birefringence*,[7,11] also called *rod* or *plate* birefringence,[2] is manifested by a *two-phased* system in which a mass of long or flat particles, *not necessarily anisotropic*, have been *oriented* in a medium of *different* refractive index by a flowing, streaming, or growing process. The slower vibration direction (higher refractive index) of the system is parallel to the length of the rods or in the plane of the plates. This kind of anisotropy is found in some fibers, films, and other colloidal systems. It is detected by a *change* in degree of birefringence *while* the specimen is mounted in a liquid (such as a standard of refractive index). Such a liquid changes the refractive index of the continuous phase, thereby changing the difference in indices between phases. Soaking in such a liquid may be sufficient to eliminate all of the form birefringence, leaving only the true double refraction. If soaking produces isotropy, all of the original anisotropy was due to form birefringence.[2]

5.   The *photoelastic effect* is the local anisotropy manifested as strain resulting from stress on normally isotropic materials such as an inorganic or organic glass[8] (Figure 5.5).

By means of the polarizing microscope, the anisotropy observed in a material can be of help in two ways by providing (a) more visibility in the microscopical *images* than can be seen with ordinary unpolarized light, and (b) *patterns* of light and dark, usually in colors. The search for both kinds of information led to the development of the petrographical microscope which carries a variety of accessories[12] added to the ordinary compound microscope. Chemists found they needed the same two kinds of information, and so the chemical microscope followed.[2] In addition to petrographical and chemical microscopes, the newer research instruments and "universal microscopes" for use in all sciences also carry polarizing equipment.

The specializations of the polarizing microscope will now be discussed in terms of the respective attributes contributing to visibility.

5.3.   NUMERICAL APERTURE AND INTERFERENCE FIGURES

The *numerical aperture*, NA, takes on special significance, particularly in the search for a principal kind of pattern, the *interference figure* (Figure 5.6). It is sometimes called a "directions image"[2] because it is

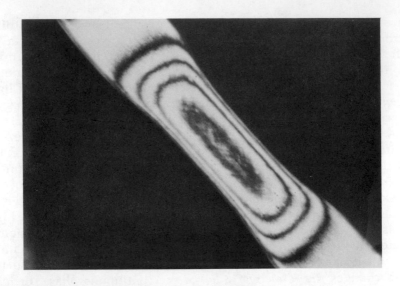

FIGURE 5.5. Photoelastic pattern in the "neck" produced by pulling a fiber of nylon by hand. (Courtesy of Bobby D. Doby, May 1973). Taken between crossed polars at 45° from position of extinction.

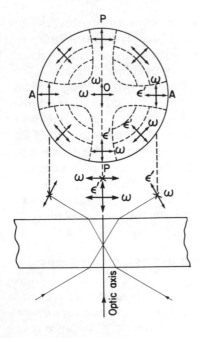

FIGURE 5.6. Diagram of an interference figure, the pattern formed by the summation of effects by the specimen on the polarized light rays coming from all the directions included in the angular apertures of objective and condenser.[2]

the pattern formed by summing up the *directional effects* of polarized light rays on the specimen, within the limitations of the *angular aperture* of the objective or its condenser, whichever is the smaller. Thus, in the production of interference patterns the value of NA transcends that for the resolution of fine detail in a pictorial image. The purpose is to obtain as large a solid cone of rays as is commensurate with the other attributes contributing to visibility, to be discussed in the usual order. For obtaining data from interference figures that are applicable to conventional data-conversion charts, some authors[4] standardize with an NA of 0.85.

5.4. RESOLUTION: INTERACTION OF SPECIMEN AND POLARIZED LIGHT

Resolution of the specimen's effects on polarized light depends on producing enough plane-polarized light with the *polarizer* and, when called for, analyzing the effects with the *analyzer*.[10] At present both polars[1] usually are made of polarizing film, e.g., Polaroid®, sandwiched between two protective pieces of thin glass. For transmitted light the polarizer P generally is placed in front of the iris diaphragm of the substage condenser, as shown in Figure 5.7.[7] The analyzer A usually is placed in a slide fitting in or out of the tube of the microscope above the objective O. In some cases the analyzer is a cap that fits over the eyepiece, but obviously such an arrangement does not provide as definite an orientation as the analyzer in a slide.

Resolution of the specimen's effects on polarized light means that everything else in the usual optical path must have no effect: Objectives, eyepieces, condensers, and microscopical slides and cover glasses must be *strain-free*. The test is to remove the specimen and, with the light source turned on and the polars crossed, look into the microscope. There should be no light. If there is a spot of light that does not fill the field, turning one optical element at a time may indicate which element is strained. If the spot does not move, remove one element at a time to find which ones are strained.[2] The manufacturer tries to supply all polarizing microscopes with strain-free components, but sometimes objectives, eyepieces, and condensers get switched around in a laboratory. The *objectives* carry some symbol such as P to indicate that they are for polarized light.

The *eyepieces* can all be of the Huygenian type because their magnifications and corrections are usually high enough for a polarizing microscope. There should be at least three eyepieces: One eyepiece should be equipped with cross lines and with a lug on the eyepiece to engage in a notch in the tube of the microscope so that the cross lines

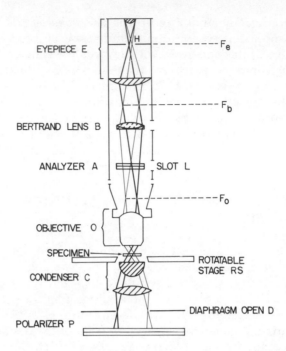

FIGURE 5.7. The polarizing microscope as used for conoscopic observations.[7] Courtesy of Microscope Publications, Ltd.

are oriented $w \overset{N}{\underset{S}{\Leftrightarrow}} E$ to the observer. Another eyepiece should be fitted with a linear micrometer (to be calibrated as instructed in Chapter 4) for distances on interference figures, etc., as discussed later. A third eyepiece should be clear of reticules for photomicrography or whenever no reticule is desired. However, other reticules, such as a protractor for estimating angles, or a cross-hatching for estimating areas, may be kept on hand, ready to drop into the eyepiece E at the focal plane F_e of Figure 5.7.

The *condenser* C should have an easily removed upper lens, such as one that flips out of the way and back into position. The iris diaphragm D is essential. The polarizer P does not need to be removable. It should be rotatable, but it should also have a catch of some sort to indicate when the direction of vibration of light through it is parallel to one of the cross lines.

Slot L in Figure 5.7 is designed to accommodate retardation plates such as a first-order red plate, or a compensator such as the quartz wedge which was used to make Figure 5.8.

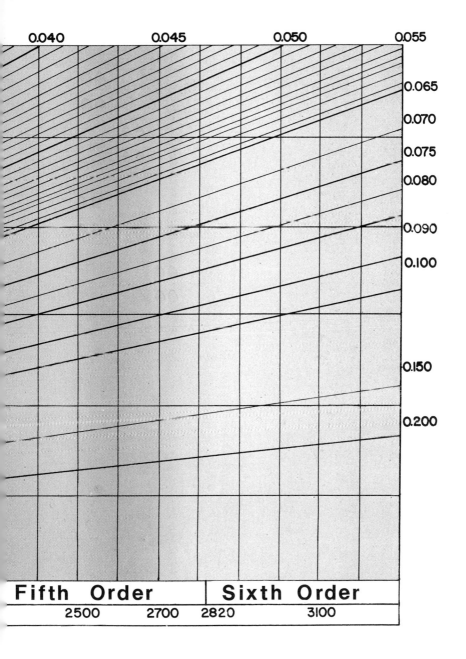

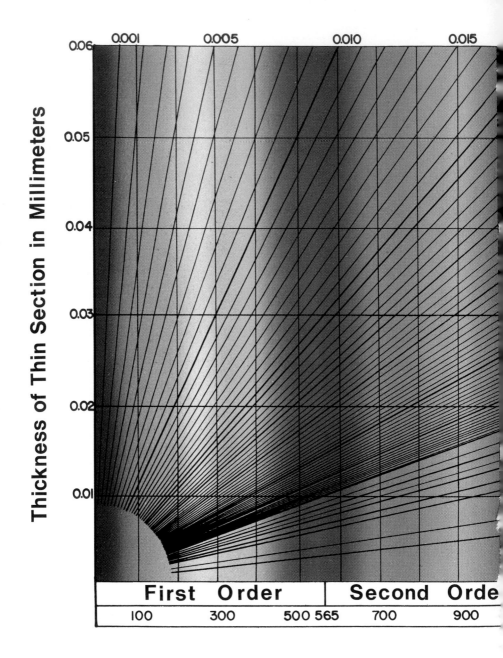

FIGURE 5.8. Michel–Levy scale of birefringence.[2,5] The color sequence (Newton's series) is based on a color photograph taken of a six-order quartz wedge placed between crossed polars. This color key is a useful approximation, but there is a slight discrepancy between the subjective appearance of the colors to the human eye and the objective rendition of the colors.

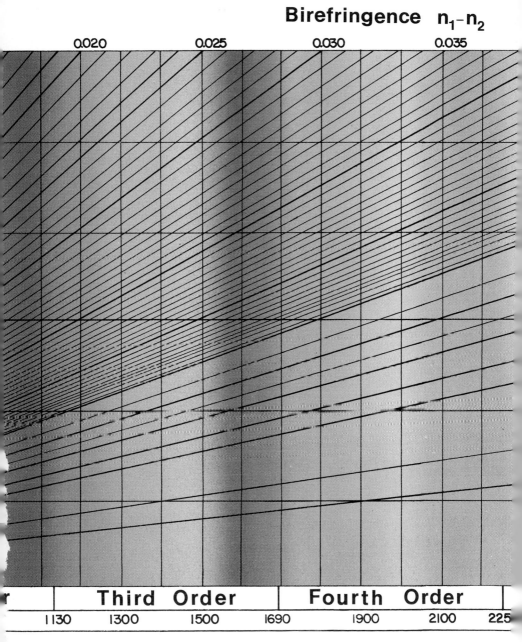

Birefringence $n_1 - n_2$

0.020   0.025   0.030   0.035

	Third Order	Fourth Order	
1130	1300	1500	1690	1900	2100	225

Retardation in Nanometers

B in Figure 5.7 indicates a Bertrand lens, a simple lens system which, with eyepiece E, constitutes a compound microscope for looking at an interference figure at the back aperture of objective O. When not in use the Bertrand lens may be slid or flipped out of the way.

Figure 5.9 shows a chemical microscope. The code of symbols is the same as for Figure 5.7.

5.5. CONTRAST: MICHEL-LÉVY INTERFERENCE CHART

Contrast, color or neutral gray on a dark background, is an outstanding attribute of viewing an anisotropic object between crossed or partially crossed polars, as has already been illustrated in Figure 2.15. The light (first-order gray or high-order white) or colors (shown in Figure 5.8) is an interference phenomenon. Figure 5.10 is a diagram used to explain how the interference comes from the functioning of the polarizer P and analyzer A. The polarizer feeds plane-polarized light to the birefringent specimen S, which splits the beam into two components.[13] The specimen is in a position of brightness, that is, it is oriented on the

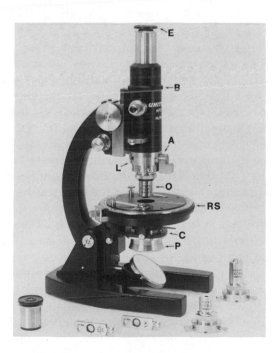

FIGURE 5.9. Polarizing microscope.

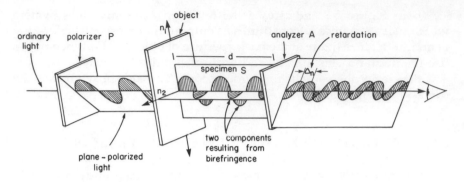

FIGURE 5.10. Diagram used to explain the function of polarizer and analyzer.[13]

rotatable stage (RS in Figures 5.7 and 5.9) so that its two kinds of rays are vibrating in different azimuths from those of P and A. Yet both bundles have to go through the analyzer A. For some wavelengths (colors) two wave trains will be in phase and reinforce each other. For other wavelengths the two trains will be out of phase and interfere with each other. The potential result is Newton's series of interference colors, depending on how much the slower component falls behind the faster component.

It is as though we have two runners, one faster than the other. Assuming that their speeds are constant, how much the slower runner falls behind the faster one depends on their difference in speed and how far they run. If we know two of the three variables (difference in speed, distance traveled, and distance between the two), we can determine the third variable. So it is with two light rays. The difference in velocity is expressed by the difference in refractive index $n_1 - n_2$. The distance the rays travel is the thickness of the specimen, expressed in micrometers, μm (formerly microns). The distance the slower light ray falls behind the faster one (retardation) is expressed in nanometers, nm (formerly millimicrons). These three variables can be put in a chart of Newton's series of interference colors, such as displayed by a quartz wedge showing several orders of retardation colors.

The classical chart is the Michel-Lévy scale of birefringence,[2] reproduced here as Figure 5.8. It was originally designed for petrographers' thin sections of rock, with the standard thickness of professional sections, 30 μm, in the middle of the ordinate. But the chart is also appropriate for fibers, since they run commercially from 10 μm to 40 μm and even greater for the thicker fibers used in carpets. The general equation for Figure 5.8 is

$$\text{retardation } (r) = \text{thickness } (t) \times \text{birefringence } (\Delta_n)^{(5)}$$

The abscissa on the bottom is graduated in nanometers of retardation. The scale is organized into orders; each order ends with red, about every 565 nm of retardation. The first order (up to $\Delta_r \approx 565$ nm), of course, starts with black ($\Delta_r = 0$) and goes from dark grays to light grays to white of the first order (1°). Some people see the dark grays as dark blues, which is just as diagnostic as dark neutral gray. The important interpretation is to recognize the grays and white of the first order. Around $\Delta_r = 275$ nm the specimen in positions between crossed polars appears to be yellow, then, at about 450 nm, it appears orange, then red, and finally magenta. The second order contains the most brilliant colors, from blues to greens, to yellows, to oranges, and finally to red. The third-order colors are less brilliant (less saturated). In the fourth order, the colors are still less brilliant, i.e., more nearly pastel. To many people the fourth-order blues seem to be faded out. By the fifth order (Figure 5.8) the blues are invisible to most people, and by the sixth order only pink (faint red) can be seen. After the eighth order, retardation is visible only as high-order white, so the sequence of orders of retardation can no longer be detected by color. We are then in need of special accessories, some of which measure as many as 30 orders of retardation. More will be said about them later in this chapter.

The upper abscissa of the Michel-Lévy chart is expressed in terms of birefrigence, $n_1 - n_2$. The scale goes horizontally from 0.001 to 0.040, and then vertically downward to 0.20. Usually the problem is to obtain the birefringence, from the thickness and retardation. The low orders of retardation are determined ordinarily with the aid of retardation plates.

5.5.1. Retardation Plates

Most polarizing microscopes are sold with two *retardation* plates as standard equipment: a first-order (1°) red plate and a quarter-wave ($\frac{1}{4}\lambda$) plate.[2] The 1° red plate is also called the sensitive-tint plate because just a very little retardation added to or subtracted from that of the red plate will turn the retardation color to a blue or a yellow, respectively. Addition of retardation means that the slower component of the specimen is parallel to that of the 1° red plate. Subtraction means that the two slower components are crossed. The direction of vibration of the slower component is marked on the 1° red plate; this provides a means for detecting the direction of vibration of the slower component as it emerges from the specimen.

Similarly, the direction of vibration of the slower component is marked on the $\frac{1}{4}\lambda$ (1° white) plate, and hence that plate can also be used to detect the direction of vibration of the slower component from a specimen with a first-order polarization color. A specimen which manifests first-order white, for example, would be dark gray or black (compensated) when its slower component is crossed with that of the $\frac{1}{4}\lambda$ plate, and bright white or yellow when the slower components are parallel. But if a white polarization color *remains white* while the specimen is rotated between crossed polars and while the $\frac{1}{4}\lambda$ plate is in place in its slot (usually NW $\leftrightarrow$ SE), the white is of high order.

Compensation is the exact matching of the retardation in the specimen with that of a standard retardation device. The criterion for matching is to observe darkness when the specimen is in a position of brightness until the retardation device renders the specimen black. Compensation would occur fortuitously with a $\frac{1}{4}\lambda$ plate if the specimen had a retardation of exactly $\frac{1}{4}\lambda$ and the two slower components were crossed. Compensation also occurs fortuitously with a 1° red plate if the specimen has a retardation of the first order and the two slower components are crossed. In either fortuitous situation, *and only then*, the retardation plate may be termed a *compensator*.

A true compensator is a device of variable retardation, such as a quartz wedge, say, of three orders. If the specimen, such as a fiber of nylon, is rotated into the NW or SE quadrant, and the quartz wedge is slid into its slot until the specimen becomes black, the compensating order and hue may be determined. From these data the approximate retardation may be estimated by using the Michel-Lévy chart. The Senarmont compensator[11] is more accurate because another graduated quartz wedge is slid over the first one until compensation is reached.

Another type of compensator uses the principle of varying the *direction* of the incident light by tilting an anisotropic material of constant thickness. One variety is called the Berek compensator.[11]

5.5.2. Thickness of the Specimen

If the exact path length is not known, it needs to be measured. Many old-fashioned and some new microscopes are equipped with *fine-focusing wheels graduated in units,* such as microns. If your microscope is one of these, you can measure the thickness of your specimen by using an objective of high NA (shallow depth of focus) and focusing upward from the *bottom* of the object. After reading the micrometer's scale with the bottom in focus, you focus upward to the *top* of the object and read the scale again. Focusing is always done in the same direction of knob-turning, so as to take up any slack or backlash in the focusing mechanism. The difference in the two readings must be multiplied by the refractive index of the specimen because the images of top and bottom are displaced by that much. Some specimens can be measured for thickness before mounting by means of a thickness micrometer.[12]

If the vertical distance cannot be measured, the distance must be measured *horizontally*. If the specimen happens to be a cube, a cylinder, or a sphere, the diameter is simply measured horizontally. Otherwise, the specimen has to be turned on its side by micromanipulation[12] or by means of a spindle stage or a universal stage.[14] Horizontal measurements are made with an eyepiece micrometer, the graduation of which is calibrated for the tube length and the particular objective in use by means of a certified stage micrometer.[12]

5.6. CORRECTION FOR ABERRATIONS DUE TO STRAIN

It is apparent by now that the most important *correction for aberrations* in the polarizing microscope is that of color. For visual work achromatic objectives are adequate, but apochromats are preferred for photography. The ordinary uncorrected condenser has such severe chromatic aberrations that its colors may be misinterpreted as retardation colors. Centering the illuminating beam and the condenser helps reduce aberrational color in an ordinary condenser. At the same time, the color temperature of the illuminating beam should be controlled by use of a daylight color filter.

All optical glass in the system of the polarizing microscope must be *strain-free*. Objectives certified as strain-free by the manufacturer are marked by a symbol, such as P. All other objectives should be tested between crossed polars. Eyepieces, condensers, slides and cover glasses, auxiliary lenses, retardation plates, and compensators should also be tested for strain; likewise windows, such as those in hot stages, should be tested for strain. (If strained, try loosening the retaining screw ring.)

Reflected light and all glare are partially polarized. One probable source of these is from overhead light entering in, or reflected from, the eyepiece. Once investigated, such aberrations are usually easy to correct.

5.7. CLEANLINESS: FREEDOM FROM INTERFERENCE FILMS

Cleanliness is especially important in working with the polarizing microscope. Thin films of oil, air, or other matter may result in interference colors that can be misinterpreted as polarization colors because they belong to Newton's series of interference colors (Figure 5.8). Care must be taken to leave no air spaces or bubbles between slide and cover glass because the appearance of the specimen in air may be misinterpreted for that in a liquid, which has a much higher refractive index than air.

5.8. DEPTH OF FOCUS

We have seen that the optical properties of an anisotropic material vary with the direction of the incident light. Theoretically, in determining these properties only axial rays should illuminate the specimen. With objectives of low NA and the top lens of the condenser flipped off, illumination is close enough to ideal and depth of focus is adequate. With high-aperture objectives, however, the iris diaphragm of the condenser should be closed as much as is feasible, and again depth of focus is a bonus. Of course resolution is decreased, and if the resolution is inadequate, fine detail should be studied separately with the iris diaphragm open.[3]

5.9. FOCUS

When anisotropic material is studied between crossed polars in a position of brightness (by rotating the stage), the material is bright against a black background, as in dark-field illumination (but for a different reason). In both instances the specimen appears to be self-luminous, like the moon at night. Such images are seldom in sharp focus. The sharpest image is obtained by first focusing with parallel polars and then crossing them to get the polarization image. Interference figures are not images; the specimen need not be in sharp focus and thick specimens usually cannot be all in focus.

5.10. ILLUMINATION

The proper *illumination* needed to obtain the optical properties mentioned in Figure 5.10 is called orthoscopic.[3] With the exception that the light is polarized, orthoscopic illumination is the axial, bright-field transmitted illumination discussed in Chapters 2 and 3. The image is the same with unpolarized or polarized light. By crossing or partially crossing the polars we get polarization images such as Figure 2.15.

When an objective of moderate or high NA is filled with light by an adequate condenser, as in Figure 5.7, the illumination is appropriately called *conoscopic*.[3] Such illumination is used primarily for observing polarization figures at the back aperture of the objective. Small but bright polarization figures are viewed with the eyepiece *out*. Larger but less brilliant figures are best seen with both the eyepiece E and Bertrand lens B *in*, as shown in Figure 5.7.[3]

With either orthoscopic or conoscopic illumination, and with the polars crossed, the incident light must be intense enough to see the effect that the specimen is producing. A Nicol polarizer, even if 100% efficient, absorbs half the light in order to polarize it. Common polarizing films absorb 65% of the light reaching them. The analyzer absorbs at least another 50–65% of the light that gets through the specimen, however transparent. So with only 17–25% of the original light available, even if the specimen fills the whole field (which it usually does not), it is seen that the initial intensity of illumination has to be high.

5.11. RADIATION

Usually a special 6-V incandescent lamp of the type already described is amply intense. Intensity is regulated by reducing the voltage as necessary by means of a variable transformer, but the radiation so obtained will be redder. A daylight filter will stabilize the color.

Something should be said here about determining the direction of vibration of the light from the polarizer, because this fact needs to be known in the detection and measurement of optical properties of the anisotropic materials discussed in Chapters 6 and 7. To date, the manufacturers of polarizing microscopes have differed about whether the $\updownarrow$ N S or the W↔E cross line represents the direction of vibration in the polarizer when it clicks into place. To determine which is actually the direction, use light reflected from a black flat surface, such as the black enameled foot of a microscope stand. As shown in Figure 5.11, *part of* the light reflected from a plane surface is polarized, and the direction of vibration lies in that surface at right angles to the reflected beam. So, if the polarizer (or analyzer, whichever is easily removable) is

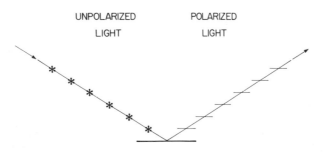

UNPOLARIZED   POLARIZED
LIGHT   LIGHT

FIGURE 5.11.   Light partially polarized by reflection from a surface.

held in the hand, looked through, and turned toward the reflected beam to maximum darkness, the direction of vibration of the light from the polar is at right angles to the direction lying in the plane of reflection.

Figure 5.11 will also explain why in reflected-light microscopy (Chapter 4) turning an analyzer will eliminate some of the glare coming from the specimen and add contrast to the image.

5.12. MAGNIFICATION

The total *magnification* of a polarization image or pattern should be no greater than is useful ($\approx$1000 × NA).[2] Empty magnification[2] only decreases intensity of the image, which may be low already.

5.13. FIELD OF VIEW OF AN INTERFERENCE FIGURE

The *field of view* of an interference figure should include only the one crystal involved, lest the figure be confused by the effects of the surrounding crystals. With small crystals, grains, or particles, a pinhole cap may be inserted in place of the eyepiece (with the Bertrand lens also removed). The smaller the hole in the cap (2 mm or less), the smaller the field of view. The Wright cross-slit diaphragm is more convenient because the size and position of its opening may be adjusted to contain only the one selected crystal observed while looking through the hinged eyepiece (replacing the regular eyepiece). When the single crystal has been framed, the Wright eyepiece may either be swung away or kept in place to work with the Bertrand lens.[2]

5.14. GLARE

With reflected illumination (Chapter 4) there is some *glare* from the specimen itself. Since a portion of any reflected light is polarized (Figure 5.11), this much of it can be removed by rotating a properly placed polar until the glare is at a minimum. As indicated before, the polar is effective also on light scattered from other surfaces, including lenses in the objective. In any case, the lower the NA of the objective, the less glare. Achromatic objectives present less glare than apochromats of the same NA because they have fewer air–glass interfaces.

5.15. CUES TO DEPTH

As already illustrated in Figure 2.15, images of anisotropic specimens obtained between crossed or partially crossed polars can be a cue to depth. The three-dimensional twists and spirals of cotton fibers present various orientations to axial polarized light so as to vary the degree of birefringence. The varying depth of the fibers changes the path length contributing to the fluctuating retardation, and consequently introduces a variety of polarization grays and colors. Figures 5.4 and 5.5 (among others) also illustrate the fact that the polarization effects are cues to depth.

5.16. DEPTH OF FIELD

In anisotropic crystals, since retardation of one of the double rays behind the other depends on the thickness of the specimen as well as the extent of birefringence, it is important that the thickness be within the acceptable focusing power of the objective. As discussed in Section 5.9, with a self-luminous object the focus cannot be as sharp as with a dark object against a bright background. Therefore the depth of field obtained with crossed polars is not as important as that obtained with bright-field illumination. What *is* important is that the thickness of the specimen be known or constant. In petrography thin sections are standardized to a thickness of 30 μm, in the middle of the range of the chart (Figure 5.8). Fortunately most fibers used for paper, textiles, and cordage are also well within the range of the chart.

Specimens for obtaining polarization figures may be even thicker than those for imaging specimens between crossed polars because polarization figures are the images of the back aperture of the objective, not of the specimen. In fact, the depth of the interference figure need be no shallower than the (long) depth of focus of the Bertrand lens (Figure 5.7).

5.17. WORKING DISTANCE

The *working distance* of an objective used on a polarizing microscope should be within the range of specimen thicknesses of Figure 5.8, plus the cover-glass thickness.[15] These are ordinary considerations for the designers of microscopical lenses. The practicing microscopist, however, should always be ready to use a hotstage[16] on his polarizing microscope, and he must have adequate working distance above it. In general, working distance is precious, and the more of it the better.

There are some extra-long working objectives designed and built primarily for universal stages.[14] These objectives are very handy for use also with hot and cold stages, microdynamometers,[17] and other setups for dynamic experiments[18] requiring objectives with extra-long working distances.

5.18. STRUCTURE OF THE SPECIMEN

The *structure of the specimen* is the most fundamental concern when employing polarized light to study a material. An anisotropic material is sure to reveal some information different from that obtained when restricted to unpolarized light. This attribute was discussed in Section 5.2, and some practical aspects of it are continued in Chapters 6 and 7.

5.19. MORPHOLOGY OF THE SPECIMEN

The *morphology of the specimen* is also important. If there is a choice of thickness, choose one that is well within the limitations of Figure 5.8. If only thicker specimens are available, think about methods of preparation that will reduce the thickness but not change the structure. If the sample is a powdery mixture of different-sized particles, optimum-sized ones can probably be screened out or otherwise separated.[18]

Crystals as received should always be looked at under the microscope even though they are likely to be aggregated, broken, or rounded. Recrystallization should always be contemplated, but employed *after* the preliminary examination.

5.20. INFORMATION ABOUT THE SPECIMEN

Information about the specimen should include history that may have brought on more or less anisotropy: stress, annealing, weathering, congealing, or precipitation due to short shelf-life, etc.

5.21. EXPERIMENTATION

Experimentation has already been discussed in Sections 5.17, 5.19, and 5.20. For example, if your information indicates a working hypothesis that will help toward solving the problem microscopically, experiment!

5.22. BEHAVIOR OF THE SPECIMEN

The *behavior of the specimen* may be planned purposely for experimentation, but it may also be observed incidentally or fortuitously. For example, both phases in a commercial sample of trisodium phosphate (TSP) were clear when received. Between crossed polars one phase was isotropic and the other phase was anisotropic. While being examined under a reading lamp, the anisotropic phase turned white (dehydrated into light-scattering pseudomorphs). This behavior called for experimentation and preparation to confirm the identity of the two phases.

5.23. PREPARATION OF THE SPECIMEN

One of the most useful methods of preparing a sample for observation with the polarizing microscope is recrystallization. Every microscopist should know how to do it well and quickly,[2] using only a few tiny crystals and a single drop of water or other solvent. For recrystallization from water, the impure or imperfect material is added to a drop of water on a corner of a microscopical slide and dissolved by warming the slide over a microheater[2] while stirring with a drawn-down tip of a slender glass rod. When all is dissolved, the slide is allowed to cool until crystals begin to form. If necessary, the slide may be warmed again to evaporate some of the water, until well-formed crystals appear. Placing a cover glass on the drop will then retard further evaporation.

5.24. PHOTOGRAPHY

For low-power *photography* with polarized light, instead of confronting the limitations of focus and light intensity of the petrographical or chemical microscope, it may be better to consider using a polarizing simple microscope[19] (Figure 5.12) or a polarizing stereomicroscope[17] (Figure 5.13). Further discussion is reserved for Chapter 8.

5.25. SUMMARY

Anisotropy arises from different speeds of light vibrating at right angles within a substance. Anisotropic long-chain molecules have un-

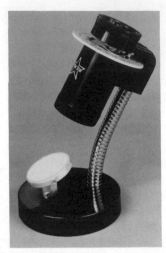

FIGURE 5.12. Simple Polarizing Microscopes.[19] These pictures are reproduced by permission of Meiji-Labax Co., Ltd., Tokyo, Japan.

FIGURE 5.13. Binocular–binobjective stereoscopic microscope with simultaneously rotatable polars.[17]

equal dipole strength, lengthwise vs. crosswise. In anisotropic crystals the directions are those of the unequal spacings of molecules, ions, or atoms, while in strained glass or plastic the directions are those of applied stress or retained strain. Form birefringence arises from differences in refractive index between long or flat particles not necessarily anisotropic, and the refractive index of another phase. Anisotropy is readily observed and measured by using polarized light, for anisotropic materials split an incident beam of polarized light into two rays ("ordinary" and "extraordinary") which do not interefere with each other because they are vibrating at right angles to each other until they are forced into a second polar, the analyzer. At some wavelengths the two emerging rays will be in phase and will reinforce each other, and at other wavelengths they will be out of phase and cancel each other. The result is a series of interference grays or colors corresponding to the regions of anisotropy. The amount of retardation of the extraordinary ray depends on the difference in refractive indices (birefringence) and the thickness of the sample, and so for a known thickness the birefringence can be estimated from the retardation color by using the Michel-Lévy chart (Figure 5.8). If the degree of interference is such that a gray or white image is obtained, the image can be shifted into a region of interference color by interposing retardation provided by a quarter-wave plate, or a first-order red plate, as indicated. A quartz wedge can provide compensation for retardation by the sample, and so provide a means of measuring it.

In practical microscopy five kinds of anisotropy arise, each very useful in identifying or characterizing a material. One kind is the anisotropy of a single anisometric crystal, the purest and the most useful because precise optical and crystallographic determinations can be made. A second kind embraces twinned crystals and other multiples, including spherulites, wherein each unit organization contributes its own characteristic optical properties. A third kind of anisotropy originates with long-chain (or flat) *molecules* having a different electronic bipolar strength lengthwise (or flatwise) than crosswise. A fourth kind of anisotropy is called form, streaming, or flow birefringence because long or flat particles can be aligned in a stream of liquid. If such particles (whether or not they are anisotropic) differ substantially in refractive index from that of the liquid, the system may be anisotropic. A fifth kind of anisotropy is the photoelastic effect, that is, strain as a result of stress. All these kinds of anisotropy can be observed and put to good analytical use by turning to microscopy with polarized light.

The polarizing microscope usually consists of a compound microscope with a rotatable stage, fitted with two *polars,* one below the substage condenser (the polarizer) and the other in the body tube above the objective (the analyzer). The polars may be of two types: the expensive but efficient Nicol prism, which transmits almost 50% of the input light as a plane-polarized beam and diverts the rest, or the inexpensive Polaroid® type, which transmits 35% of the incident light in polarized mode and absorbs the rest. One polar is rotated until its vibration direction is at right angles to that of the other polar, so that the field is dark and any anisotropy in the specimen will appear luminous. By using an appropriate retardation plate, the sample and its background will appear in contrasting colors.

It is evident that the objective, eyepiece, and condenser of a polarizing microscope must all be free of strain so that they do not introduce extraneous anisotropy. High NA is desirable in the condenser and objective, not only for resolving power but also to include a large-enough angular aperture to develop *interference figures*, which are summations of all the directional effects of polarized light in the specimen within the solid cone of light.

A polarizing microscope should have its fine-focusing mechanism graduated in microns, so that the thickness of a specimen can be measured by focusing first on the bottom surface and then on the top surface. The use of different types of compensators to measure the degree of retardation was discussed in Section 5.5.1.

Two kinds of illumination are important in microscopy with polarized light: orthoscopic and conoscopic. *Orthoscopic* illumination is that which is ordinarily used to form a bright-field image of the object, as discussed in Chapters 2 and 3; also, with polarized light the condenser's diaphragm is partially closed. In *conoscopic* illunination the condenser's aperture is set to equal that of the objective (usually with $NA = 0.85$). Instead of viewing the image of the object, the back aperture of the objective is viewed by removing the eyepiece or by inserting the Bertrand lens between objective and eyepiece. To obtain the interference figure of only one crystal at a time (since neighboring crystals are probably oriented differently), a pinhole cap or a cross-slit diaphragm is necessary when the field of view contains more than the one crystal.

Depth of field is important in polarized-light microscopy because total birefringence depends on thickness of the sample and because standard petrographic sections are 30 μm thick and textile fibers are typically 10–40 μm thick. Working distance is also a consideration because the microscopist usually wants to conduct experiments on the sample, and often needs to use a tilting stage, hot stage, or cold stage. To carry out such experiments he must know how to prepare the specimen, making it thin enough without changing its structure, and how to recrystallize material well on the microscope slide.

The Microscopical Properties of Fibers

6.1. INTRODUCTION

Fibers are unique units[1] of biological tissues, mineral habits, and spinning processes. Examples are muscle and nerve fibers, wool, cotton, linen, natural silk, natural and regenerated cellulose, asbestos, fine wire, spun silicate glass, and man-made polymeric fibers. While they vary widely in chemical nature, fibers are physically alike in being very much longer than they are wide,[2] very strong for their small cross sections, and optically anisotropic. The kind and extent of optical anisotropy vary greatly among species of fibers according to the five classes defined in Section 5.2: single or spherulitic crystals, molecular birefringence, form birefringence, and photoelastic effect. Consequently the optical properties of fibers[3,4] are discussed separately from crystals.[5,6] Fibers do not display crystal faces, so there are fewer optical properties to be observed. Furthermore, variations in composition and treatment[7] produce variations in the optical values of a given species of fiber,[3] whereas constant values are the rule for a definite species of crystal.[5,6]

About eight optical properties of fibers may be observed. No previous knowledge of crystallography is required. We can begin with Figure 6.1 by considering a simple fiber in which light waves vibrate in two privileged[5] directions: one vibration direction parallel ($\parallel$) to the fiber's axis, and the other vibration direction perpendicular ($\perp$) to the axis. In order to consider the two directions simultaneously, a few fibers are cut to lengths of 1 or 2 cm and mounted in a liquid to reduce the scattering of light by the fiber in air (refractive index $n = 1.00$). Water ($n = 1.33$) is handy for mounting hydrophilic fibers such as cotton, and mineral oil ($n = 1.48$) is commonly used for hydrophobic fibers such as nylon. The fibers should be examined immediately upon being mounted in the

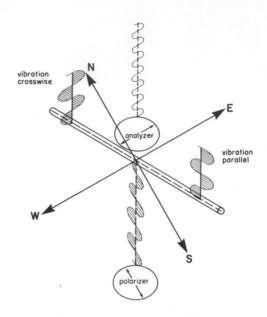

FIGURE 6.1. Fiber in position of brightness, between crossed polars. Drawn by C. Miller.

liquid, because they may change. Cotton, for example, swells in water with corresponding changes in optical properties, and nylon swells in some organic liquids. Such changes should be noted as behavioral information, but the initial characteristics are to be compared with determinative literature.[3] At the same time, subsequent behavior in liquids may indicate that more experiments may be worthwhile.

With the fibers properly mounted, the eight determinative optical properties may be observed in succession with the aid of a polarizing microscope. The fibers are first brought into focus with the analyzer out, and then are observed between crossed polars while the stage is rotated slowly. The eight points to be noted are given below, with an explanation of each.

6.2. THE EIGHT OPTICAL PROPERTIES OF FIBERS

1. *Brightness* or grayness (instead of darkness) in some or all positions of rotation *between crossed polars*. Most types of fibers show brightness in only certain positions of rotation (Figure 6.2), but cotton is bright in all positions because its structure is spiral instead of axial.[6] Inorganic glass fibers appear to show no change when rotated between

crossed polars. Glass fibers are too thin to show detectable retardation ordinarily, although thicker glass rods and bottles do. The fact that glass fibers appear to be isotropic, even though they are not strictly so, should not detract from the analytical importance of their being the only kind of fiber that comes so close to showing no retardation under the microscopical conditions set forth above. That alone is determinative.

2. *Extinction* (darkness), complete or incomplete, every 90° of rotation between crossed polars. (See Figure 6.2.) Cotton, with its spiral structure, shows no substantial extinction. Linen (flax), hemp, certain other bast fibers, and some mechanically treated man-made fibers show extinction in certain segments at a time.

3. The extinction may be *parallel or oblique* to the axis of the fiber, and the angle is determinative. Some fibers display segmented extinction, as in bast fibers, silk filaments, and some crimped man-made fibers (Figure 6.2).

4. *Retardation*, the distance in nanometers that one wave train falls behind the other (Figure 6.1), is manifested as grays, whites, and colors (Figure 5.8), some of which (reds and pinks) are visible between crossed polars to the eighth order of classification of the interference between slower and faster ray-bundles. Some polyester fibers show pinks of such high order.

5. *Sign of the birefringence*, depending on which of the two ray-bundles is the slower. Most kinds of fibers are positive, signifying that the rays vibrating lengthwise are the slower (i.e., have the higher refractive index, n). But an important minority of types is negative. The sign is determined between crossed polars using a *retardation plate*.

6. *Quantitative birefringence*,[2] the numerical difference between the high and low refractive indices, is independent of path length (thickness, denier). It may be determined from the retardation per unit

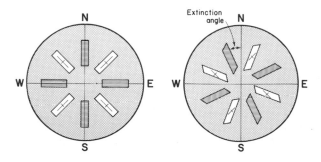

FIGURE 6.2.   Positions between crossed polars of brightness and extinction (darkness). (left) parallel extinction, (right) oblique extinction.[6]

path length, by means of *compensators*[8-11] (see Chapter 5). Birefringence may also be determined *at the surface* of the fiber by measuring the two separate refractive indices by immersion techniques.

7. $n_\parallel$, the refractive index for light (from the polarizer) *vibrating parallel* to the length of the fiber, may be measured by comparing $n_\parallel$ with the refractive index of a liquid standard, as will be explained later. The resulting datum is determinative.

8. $n_\perp$, the refractive index for the light *vibrating perpendicularly* to the length of the fiber, is measured in the same way as $n_\parallel$, but with the fiber oriented at right angles to the position used for measuring $n_\parallel$. This datum is determinative, and may also be used to determine the quantitative birefringence as in property 6, above.

Some explanation and further comment about each of the eight optical properties of fibers is now in order, in the same succession:

1. *Brightness* results when an anisotropic fiber on the rotatable stage is so oriented between crossed polars that there is a vector of each direction of vibration parallel to the direction of vibrating light rays in the analyzer (Figure 6.1). The vector varies from nothing (darkness) at positions of extinction to maximum brightness at a position of 45° between extinctions. When there is no extinction, as in the case of a spiral structure such as that of cotton, there is always brightness between crossed polars. In most kinds of fibers the brightness is strong enough for the microscopist to recognize with confidence that the fiber is anisotropic. In a few cases, such as with some cellulose acetates and some acrylics, the "brightness" may be so weak as to leave doubt of anisotropy. In fact, some cases of *iso*tropy have been reported in the literature for some trademarked products of these generic types. When in doubt, you should bring the sensitive tint of a first-order red plate into play. Also, put a daylight blue filter into the path of yellowish artificial light. Now the field will be bright red of the first order. If the fiber *is at all anisotropic*, in one diagonal set of quadrants, the fiber will be blue, or at least magenta. In the opposite set, the fiber will be yellow, or at least orange (Figure 6.3). No further proof of anisotropy is needed.

2. *Complete extinction* (darkness) means that the whole fiber extinguishes between crossed polars every 90° of rotation. Occasionally, natural silk fibers show some short portions which do not extinguish in the same positions of rotation as do the major portions. Characteristically, these positions are where the still-soft (recently spun) filament was crossed by another part of the same filament as the silkworm was criss-crossing its double filament to prepare its cocoon. Flax (linen) and other bast fibers are really *bundles* of fibrils. Like a bundle of straws when bent, some of the fibrils will be dislocated. Each dislocation results in a local relocation of extinction, which shows up as a short, cross-hatched sec-

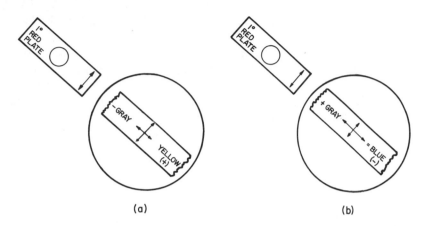

FIGURE 6.3. Determination of the sign of birefringence of a fiber manifesting a first-order gray.

tion. The number of dislocations indicates the number of bendings. Thus, old linen may be distinguished from new linen, and linen-rag paper may also be recognized.

3. *Parallel extinction* means that the axis of the fiber is parallel (or very closely parallel) to one of the two cross lines in the eyepiece, when the fiber becomes dark during rotation of the circular stage between crossed polars. It is assumed that the cross lines have been tested with a standard specimen of parallel extinction so that one cross line in the eyepiece represents the direction of vibration of light from the polarizer $\left(\text{e.g.,} \begin{smallmatrix} N \\ \updownarrow \\ S \end{smallmatrix}\right)$ and the other cross line represents the direction in the analyzer (e.g., w↔e). (See Figure 6.2.)

Parallel extinction is shown by practically all man-made fibers, *as spun*, because the structural units are arranged to the direction of flow through the spinnerette. Treatments, such as crimping or texturizing, locally alter parallel extinction if the parallel arrangement is disturbed.

4. *Retardation* of one of two wave trains behind the other is the total effect on the particular fiber by polarized light. In Figure 6.1, the polarizer is set so that the direction of vibration of the emerging light waves is that of north–south, applying terms of a magnetic compass to the cross lines of the eyepiece. The fiber is either oriented NW\SE or SW/NE because (having parallel extinction) it has maximum brightness in these two positions when the analyzer is crossed (w↔e) with the polarizer $\left(\begin{smallmatrix} N \\ \updownarrow \\ S \end{smallmatrix}\right)$. Referring to Figure 6.1, the birefringent fiber splits the

polarized beam into two polarized rays, one vibrating lengthwise (‖) of the fiber and the other crosswise (⊥). They travel through the fiber at different velocities without interfering with each other because (as shown in Figure 6.1) they are in different planes. Yet they must squeeze through the analyzer, which is set for vibration in a single direction (W ↔ E). Two rays, one from each wave train, each with the same wavelength, may have traveled through a fiber of just the right thickness (denier) to enter the analyzer in such phase that they completely cancel each other. That is, the color corresponding to that wavelength is lost in the analyzer. But at some other wavelength the rays will be in such phase that in going through the analyzer they reinforce each other. If, for example, the difference in velocity between the rays vibrating ‖ and those vibrating ⊥ is such, and the thickness of the fiber is such that the slower ray is in phase but about 525 nm behind the faster ray, the color of the light emerging from the analyzer will be red of the first order, as shown in the Michel-Lévy scale (Figure 5.8). If the retardation is about 1050 nm, the color is red of the second order; about 1600 nm, third-order red; about 2100 nm, fourth order, and so on. Intermediate retardations are manifested in the other colors (at least through four orders) in Figure 5.8. The second-order colors are the brightest in this (Newton's) series of interference colors. In the fourth order the blues and yellows have faded out. In higher orders than those shown in Figure 5.8, all colors but red have faded and even the red is lightened to pink.

The *order* of retardation colors may be difficult to recognize. In such cases a uniform quartz wedge of three or more regularly increasing orders is useful. The wedge is placed in the slot in the body tube. If this slot is in the NW\SE diagonal, start the test with the fiber oriented in the same direction. If, as the wedge is gradually inserted, the colors in the fiber go up in order, rotate the fiber 90° to the other direction. Now, as the colors go down in order, count the orders to compensate (to darkness) the color band(s) in question.

Drawn polyester fibers ordinarily exhibit orders higher than the three or four orders of modestly priced quartz wedges. However, higher orders may not be needed in this particular test if the third or fourth red band can be recognized so that the higher-order reds (pinks) can be counted in the *thickest part* of the fiber.

Besides the bands of different retardation colors due to differences in fiber thickness, there are those due to differences in birefringence, such as in the skin-and-core structure of some fibers.[7] These and other radial heterogeneities may be detected in cross sections, if they are not recognized in the longitudinal views.

Retardation (polarization) colors and their order of extent (Figure

5.8) are very determinative, once recognized. Granted that the thickness varies widely among natural fibers, and purposely in different deniers of man-made fibers, the variation is not too great to spoil the general value of retardation colors and their orders. Low-order grays and colors are especially significant because variations in thickness (denier) make so little difference in the order of retardation. Acrylics and cellulose acetates, from fine deniers to carpeting sizes, always show first-order grays or colors, whereas practically all drawn polyesters manifest high polarization colors. The difference in order of colors between undrawn and drawn nylon and polypropylene is so great that degrees of drawing may be recognized. Immature cotton fibers show yellow or blue on the red blackground of a first-order red retardation plate. On the other hand, mature cotton fibers are distinguished by colors of slightly higher order.[6] Recognition of distinctive retardation colors in a mixture of two or more kinds of fibers is not only useful in forming quick qualitative opinions but also in estimating the proportions of those kinds of fibers.

5. "Sign of birefringence" (or retardation) is preferred over the term "sign of elongation," since the sign of birefringence pertains to all anisotropic fibers, whether or not they have been elongated (stretched). Most fibers are positive (+).[3] That is, they are length-slow: $n_{\parallel} > n_{\perp}$. But some are negative (−), i.e., length-fast: $n_{\parallel} < n_{\perp}$. Types with negative sign at room temperature[10] are acrylics, most cellulose *tri*acetates, and at least some sarans and vinyons.[3] The fact that cellulose triacetate has a negative sign, while cellulose diacetate has a positive one, indicates the ability of the acetyl side group to slow down light and shows the best way to differentiate between the two degrees of acetylation.

To determine the sign of fibers showing first-order *gray*, a first-order red plate is inserted while the fiber is in a position of maximum brightness. When the slot for the retardation plate is oriented "northwest–southeast" (NW\SE), the fiber also is oriented NW\SE (Figure 6.3); otherwise, SW/NE for both. If the sign of birefringence is positive (e.g., cellulose diacetate), the fiber will be yellow or orange; if the sign is negative (e.g., cellulose triacetate), the fiber will be blue or magenta. Gray on the Michel-Lévy chart (Figure 5.8) amounts to a retardation, say, of 100 nm. The retardation of the first-order red plate is about 500 nm, and its direction of slow vibration is ordinarily oriented crosswise to the long plate holder (Figure 6.3). Thus, 500 − 100 = 400 (yellow); 500 + 100 = 600 (blue).

In a fiber of *positive* birefringence the slower ray vibrates parallel to the fiber's axis. According to one convention,[6] the direction of vibration for the *slower* ray is represented by an arrow shorter than the one representing the direction of vibration of the faster ray. As shown in Figure 6.3a, in the diagram of a positive (+) fiber, the shorter of the two perpendicular arrows is drawn *parallel* to the fiber's axis and is crossed with the (short) arrow of the first-order red plate. The situation may be recalled by considering that the two short arrows make a plus (+) sign, signifying a positive (+) sign of birefringence for the fiber. All this is deduced from the observation that the fiber *alone* was *gray* between crossed polars and became *yellow* when the 1° red plate was inserted. Since yellow represents a lower retardation than the 1° red of the plate, there is a *subtraction* (reduction) in retardation of the 1° red plate by the fiber when in the NW\SE position. Hence, the direction of vibration of the slower rays of the fiber must have been crossed with that of the retardation plate; the slower rays must be *lengthwise* and the fiber must be positive.

In the other case (Figure 6.3b), the fiber happens also to be gray in position of brightness between crossed polars, but it turns *blue* in the NW\SE position. Blue represents a *higher* retardation than the red of the plate. That is, the two slow-ray directions of vibration are *parallel*, indicating a negative sign of birefringence.

In some cases of skin-and-core construction,[12,13] such as in some acrylics, the core has a different sign (+) than the skin (−).

With any retardation plate (or compensator), be sure of the direction of vibration of the slower ray. If there is only one arrow engraved on the mount of the plate, it represents the direction of vibration of the slower component. But some manufacturers may choose to have that direction oriented lengthwise to the plate's metal mounting; or, if the markings are $\uparrow x' \rightarrow z'$, the direction z' is that of the slower ray. Since the retardation disk itself *may become turned in its mount*, check with a known material such as cellulose diacetate (+), or with any acrylic (−) except Orlon® (Orlon® may be too weakly negative to use as a strong standard). Be careful in cleaning any optical disk mounted in a cylindrical well; do not use circular pressure sufficient to turn the disk. If this should happen, you or a repairman will have to loosen the pressure cap and turn the disk back. Always make sure the pressure cap is tight.

The first-order red retardation (sensitive tint) plate is good for determining the sign of birefringence of any material which shows a retardation (polarization) gray or color (yellow, orange, red) of the first order. The same plate will also indicate certain colors of the second order, such as blue or green, by providing a dark or light gray during subtraction (crossing of slow directions) and a blue or green of third order during addition (parallel slow directions). But for yellow, orange, or red of the second order, subtraction or addition of first-order red will merely give the corresponding color of the first or third order. These may be somewhat difficult to tell apart.

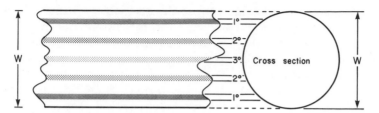

FIGURE 6.4. Diagram of a cylindrical fiber, showing a portion of the longitudinal view, between crossed polars. Hypothetically, the birefringence and thickness (width, w) are such as to exhibit three orders of interference colors (1°, 2°, 3°).

The *quarter-wave ($\frac{1}{4}\lambda$) plate* (light gray) is also good for determining the sign of birefringence of a material showing a first-order *gray*. Subtraction of the two retardations will give a darker gray; addition will give a brighter gray (or a yellow). With a color of low order the $\frac{1}{4}\lambda$ plate gives a *different* color for subtraction than for addition.

For fibers showing retardation *colors* (see Figure 6.4), the *quartz wedge* may be used as a retardation plate by *moving* it between crossed polars over the fiber in a position of brightness. *Subtraction* results in a succession of colors *downward* in order.[6] *Addition* results in an *upward* succession of colors. With knowledge of the direction of the wedge's slower component, the sign of the fiber's birefringence is determined as indicated by the arrows (not the colors) in Figure 6.3.

Dichroism is the preferential absorption of all* or part of the spectrum in one preferred direction relative to the other preferred direction. The phenomenon is visible especially well with deeply dyed fibers. Since dye in the fibers obscures the complementary polarization color, the phenomenon of dichroism can substitute its own evidence of anistropy. Usually the absorption is greater in the direction in which the slower ray vibrates. In positive fibers this direction is parallel to the axis of the fiber; in negative fibers it is perpendicular. Therefore, dichroism can indicate the sign of retardation (birefringence). The indication can, of course, be confirmed by the use of retardation plates with the fiber in a diagonal position (instead of parallel) to the cross lines of the eyepiece.

6. The arithmetic degree of *birefringence* is the difference between the two principal refractive indices, e.g., $n_\| - n_\perp$.[11] Indeed, both

---

* It is interesting to note that dichroism is the fundamental mechanism in polarizing films, such as Polaroid®. In the colorless variety, practically all of the visible spectrum is absorbed in one preferred direction but transmitted in the other preferred direction.

the birefringence and its sign can be determined by measuring $n_{\parallel}$ and $n_{\perp}$ and subtracting them. However, the measurements may be long and tedious. Moreover, they are determined only at the surface, which may be different in refractive index from the interior, as in the case of skin-and-core construction.[7,13] Such radial heterogeneities are actually recognized by the *variations* in birefringence. Therefore the degree of birefringence is a separate and independent optical property and a very important one.

Birefringence alone allows for a broad classification[3] of fibrous types into those of *weak* birefringence, 0.001–0.01 (e.g., natural and regenerated proteins, acrylics, modacrylics, acetates, saran, vinyon); those of *moderate* to *strong* birefringence, 0.01–0.1 (e.g., natural and regenerated cellulose, silk, nylons, polyolefins, vinyl); and those of *intense* birefringence, above 0.1, as in polyesters.[3]

Birefringence can be defined in other terms as the ratio between retardation and thickness of the specimen at a given point. Indeed, the quickest and simplest way to estimate the birefringence is from the *retardation* color and the *thickness* of the fiber, employing the Michel-Lévy scale of birefringence.[6] If the fiber is practically a cylinder, as is the case for some nylons, polyolefins, polyesters, rayons, acrylics, and modacrylics, the thickness is the same as the *width* of the longitudinal view and therefore can be measured directly with a calibrated ocular scale, as explained in Chapter 5. Fibers of weak birefringence, such as acrylics and modacrylics, usually display first-order gray between crossed polars. Note particularly the shade of gray along the axis of the fiber and match this gray with the likeness on the Michel-Lévy scale. Go up the ordinate of the scale until the thickness of the fiber is reached; then follow the *diagonal* line to the birefringence and read the result.

Fibers with moderate to intense birefringence and the usual parallel extinction display color bands parallel to the axis. The number of bands (degree of retardation) depends on the fiber's birefringence and path length. With a circular cross section,[3] as shown in Figure 6.4, there is a single central band of colors, and in this particular case it is flanked by two pairs of color bands of descending order. The colored line that is congruent with the axis of the fiber represents the highest retardation of all and is the value, together with the diameter of the fiber, that is used with cylindrical fibers in the translation into birefringence by means of the Michel-Lévy scale.

If the cross section is not circular, or even roughly so, it should be studied so as to anticipate where the color band(s) of highest order will appear in the longitudinal view of the fiber. For example, in Figure 6.5 the cross section is dumbbell-shaped.[3] Such a fiber will most probably lie on its side, which is then seen as the width (W) in longitudinal view

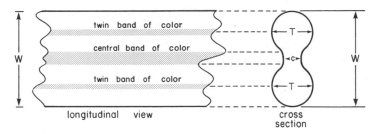

FIGURE 6.5. Diagram of a hypothetical fiber with cross section like a dumbbell and, hypothetically, with one set of twin color bands. Considering the shape of the cross section, the twin color bands must be of higher order than the single central band.

and which is measured on the particular fiber in view. The thickness of the particular fiber is estimated from the ratio of the thickest (T) part or the center (c) to the width (W) of a typical cross section of the sample. Then, the retardation colors of either the central band or the twin bands are compared with the respective thicknesses to estimate the bire-fringence from the Michel-Lévy chart (Figure 5.8).

Likewise, the cross sections of fibers of some other consistent shapes, such as dogbone, bean, cogwheel, and triangle (trilobal) can be used to estimate what proportion of the width (measured in longitudi-nal view) is the thickness with respect to a particular retardation color, so as to estimate birefringence from the chart. Fibers with tubes are often regular in cross section, and so thickness in longitudinal view can be estimated from the measured width. Even irregular cross sections, such as are commonly found in cotton, some viscose and acetate fibers, and some dual-component fibers, may give enough information about the ratio of thickness to width.

The thickness of a fiber can also be measured directly, if means are available for turning or twisting it so as to measure it on its edge.[7] Or possibly a microscope is available with a graduated, calibrated fine-adjustment focusing wheel for measurement of the vertical distance from bottom to top,[6] as explained in Chapter 5. Be sure to multiply the apparent distance by the average refractive index of the fiber.[6] A very rough estimate of the average thickness may be made from the relation

$$\text{denier} = \pi d^2/4 \times 9000 \times \text{density (or sp. gr.)}$$

Thickness, d, is in millimeters in the above formula and in the Michel-Lévy chart (Figure 5.8). A still rougher estimate can be made by remem-bering that most fabric-making fibers (except those for carpeting) are between 0.01 and 0.03 mm in average diameter.

The Michel-Lévy scale is inadequate for polyesters that manifest retardation colors of a higher order than six. For such fibers a high-order compensator is required.[8] This brings up the general subject of measuring birefringence by means of a variable retardation device, the *compensator*.[11] A quartz wedge may be used directly if it is of a sufficient number of orders and if it is graduated. The Babinet compensator is more elaborate; it consists of two opposing quartz wedges. One of them is moved over the other by means of a graduated micrometer screw.[9] The Berek type of compensator employs a different principle, that of measuring the degree of tilt of an anisotropic plate to the point of compensation.[11] Berek compensators are available in various ranges. For highly drawn polyester fibers a range of 10 orders is needed, and for experimental fibers 15 or more orders may be called for.[12]

Another important point is that exact compensation in white light can be achieved only if the dispersion of birefringence of the fiber and that of the compensator are the same.[12] For path differences under four orders no difficulty is found, but with higher orders the band of compensation is no longer black but rather unsymmetrically colored and broad. One solution is to establish the color of the compensation band. Another solution is to cut a wedge out of the end of a fiber and examine it in a subtractive position between crossed polars with the Berek compensator in place. A count of the resulting fringes plus any additional fraction gives the retardation.[12] The thickness in millimeters still has to be determined or estimated as described above.

The determination of birefringence has many applications in production and in quality control as well as in research. More and more, the determination of birefringence is being used to control the degree of elongation for optimum strength and for other desirable properties.

7 and 8.   While birefringence alone may be a key to understanding the relationship between elongation and strength, it may also be important to know the actual individual values of the two principal refractive indices of a fiber. The index for light vibrating parallel to the fiber's axis, $n_\parallel$, varies from a low of about 1.47 in cellulose acetate to as high as 1.73 in some polyesters. The span for $n_\perp$ is not as great, but is large enough to give a good spread among types. It varies from about 1.47 in acetates to about 1.54 in polyesters and about 1.61 in saran.[3]

The determination of refractive index is generally made by trial and error, and is usually tedious. But experience, technique, cleanliness, and care can make the results very satisfying. In order to separate $n_\parallel$ from $n_\perp$, polarized light is supplied by the polarizer only. *The other polar is removed from action.* Figure 6.6 illustrates the orientations of the fiber for determining $n_\parallel$ and $n_\perp$, respectively, with the polarizer having its direction of vibration $\updownarrow$, as represented by the respective cross line of

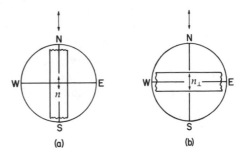

FIGURE 6.6. The alignment of a fiber to determine (a) $n_\parallel$ and (b) $n_\perp$ for a fiber with parallel extinction and positive birefringence. The direction of vibration of the polarized light is $\updownarrow$.

the eyepiece. In Figure 6.6a the fiber is aligned parallel to the $\updownarrow$ cross line, so that $n_\parallel$ is being measured. In Figure 6.6b the fiber is aligned parallel to the w↔e cross line, so that $n_\perp$ is being measured. Incidentally, the shorter arrow is shown lengthwise, that is, $n_\parallel$ is in this instance greater than $n_\perp$, and the fiber has positive birefringence.

There are two ordinary methods (among special procedures[14]) for determining the refractive indices of a fiber. In both of them the fiber is immersed successively in liquids of known refractive index, and a search is made for a match between liquid and fiber for $n_\parallel$ and for $n_\perp$. One method for determining whether the index of the fiber is higher or lower than that of the liquid makes use of oblique illumination, and the criterion is the position of the resultant shadow cast by the specimen.[6] The other method is that of the Becke refraction line and its critical movement during the process of focusing. The Becke method is preferred when the refractive index of the liquid is close to that of the fiber. In either case, the specifications for illumination are very important, and a series of liquids of known refractive index is required.[15,16]

The Becke test requires axial transmitted illumination. If the condenser is in place, *remove the top lens* and nearly *close the iris diaphragm*, leaving the polarizer in place. Focus up and down and notice the movement of the Becke halo around the fiber. *As you focus upward the halo moves to the medium of higher refractive index* (Figure 6.7). When you are far from a match between fiber and liquid,

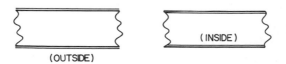

(OUTSIDE)   (INSIDE)

FIGURE 6.7. The Becke line. Simulated fiber is in a liquid of higher index: (left) on focusing upward slightly, (right) on focusing downward slightly.

the over- or under-match is very evident because the halo is in bold contrast. As you approach a match the halo becomes fainter. Darken the room if it helps. Concentrate. Be patient. The Becke method is very satisfactory to the experienced observer, and of course experience always helps. From consideration of the first six optical properties and other information, you should have a good idea of what your specimen is and what refractive indices are expected. Choose for your first liquid one with a refractive index in the middle of the probabilities. Learn to estimate how much over or under the correct index you are; then take a bold step, trying to bracket the index as soon as possible. Your previous data, especially regarding the sign of birefringence (which is greater, $n_{\parallel}$ or $n_{\perp}$?), and the amount of birefringence (difference between $n_{\parallel}$ and $n_{\perp}$), will help considerably in your choice of trial liquids. Use short lengths of each of a few fibers to avoid entanglements and bending. Be careful not to contaminate your specimens with any other kind of fibers. Use small cover glasses to save standard liquids and still be able to fill the cover glass and slide. When no Becke halo is observable and nothing seems to move as you focus up and down, the fiber and liquid have the same refractive index. Usually, however, the refractive index is an intermediate between the values of those of two successive liquids in the series. Repeat with the fiber in the other orientation, and you have both $n_{\parallel}$ and $n_{\perp}$.

The *oblique illumination* test is advantageous when the specific refractive index of the fiber is quite different from that of the test liquid. This situation can happen with an unknown fiber. The question is whether the fiber is of much higher or lower index than the liquid. Sometimes it is difficult to tell whether the bold, dark Becke line is moving in or out during focusing, whereas it is easy to tell whether a dark shadow is on one side or the other of the bright fiber illuminated *obliquely* across it. To accomplish this, the *full* condenser is adjusted in focus with its *diaphragm wide open*. The fiber is oriented $\overset{N}{\underset{S}{\updownarrow}}$ or W↔E, depending on which of the two indices is to be determined. Then a long piece of black cardboard is placed to one side underneath the condenser with its long edge $\overset{N}{\underset{S}{\updownarrow}}$ or W↔E, depending again on which index is being determined. This will give oblique illumination, with half a cone of light. If the fiber is of *higher* refractive index than the standard liquid, its dark shadow will be on the same side as the dark side of the field (Figure 6.8). If the fiber is of *lower* index than the liquid, the shadow will be on the side of the fiber which is opposite the dark side of the field.

Sometimes it is important in research to determine $n_{\parallel}$ and $n_{\perp}$ for a

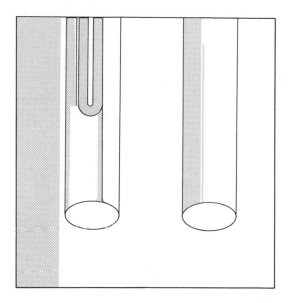

FIGURE 6.8. Oblique transmitted illumination. Glass tube and rod, in a liquid of lower refractive index, which has partially filled the tube.

series of samples, so as to compare variations in index that result from some variation in physical condition, such as percent elongation. Here the birefringence may not vary much, but the precise values of refractive index may change in some characteristic manner. For example, some acrylic fibers show an increase in both $n_{\parallel}$ and $n_{\perp}$ when they are stretched, and even though the difference between the two indices (the degree of birefringence) does not change noticeably, the magnitudes of the indices indicate at once that the fibers have been stretched.

The microscopy of fibers is not yet a highly advanced science, but in the relatively short time it has been applied to the technology of textile fibers it has proven to be of great analytical value.[17] It allows the microscopist to distinguish easily between natural and man-made fibers.[18] By comparing the results with the published values for the common classes of fibers, some very useful relationships become apparent.[19–21] The properties are found to stem from (1) the chemical (molecular) composition of the fiber, as was to be expected, and (2) the changes in orientation and spacing of the constituent molecules or other structural units during spinning, stretching, or other treatment. This book is not a treatise on fiber chemistry or technology, but some generalizations about structure and composition vs. properties will help the microscopist to understand how anisotropy comes about, and what governs it.

6.3. MOLECULAR ANISOTROPY

Molecular anisotropy originates with electric polarity between adjacent atoms. Figure 6.9 illustrates the three simplest examples:

a. Figure 6.9a shows a pair of adjacent atoms aligned with the direction of vibration of the polarized light.

b. Figure 6.9b shows a pair of adjacent atoms aligned at right angles to the vibration direction.

c. Figure 6.9c shows two atoms relatively far apart.[16]

In all three cases the polarized light waves tend to orient the electrons and nuclei of the constituent atoms in the direction of the vibration of the polarized light $\left(\text{e.g., } \begin{smallmatrix} N \\ \uparrow \\ S \end{smallmatrix}\right)$, producing *electric dipoles*. The three kinds of dipole alignments are of different strengths. Each of the *adjacent* dipoles aligned parallel to the vibration direction is rendered stronger, and each of the adjacent dipoles aligned perpendicularly is rendered weaker than each of the separated dipoles shown in Figure 6.9c. In long molecules such as those of cellulose, nylons, polyolefins, and polyesters, atoms are largely chained along the length of the molecule, so that the dipoles are strongest when light is vibrating in the lengthwise direction of the molecule.

The same molecular birefringence is imparted characteristically to fibers, as suggested in Figure 6.10. The degree with which the long

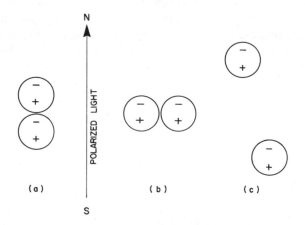

FIGURE 6.9.   Effect of polarized light waves on atoms and atom pairs.[16]

FIBER   AXIS

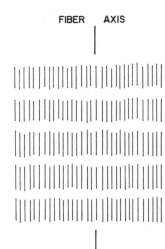

FIGURE 6.10. Idealized alignment of long-chain
molecules parallel to the axis of a fiber.[16]

molecules grow or are made parallel to the axis depends on conditions of
formation[10,13] and the subsequent treatment.[17]

With straight-chain molecules substantially like those shown in
Figure 6.10, the refractive index (which is a positive function of the
strength of the atomic dipoles) is greater for light vibrating parallel to
the fiber axis than that for light vibrating perpendicularly to it. That is,
$n_{\parallel} > n_{\perp}$ and the fiber is optically positive. For example, polyesters are
intensely positive; cellulose, nylons, and polyolefins are strongly to
moderately positive, but acrylics and cellulose diacetate are only weakly
positive[3] because of the opposing polarity of the side groups on each
molecule. In fact, the third acetyl group of cellulose triacetate is of sufficient counterpolarity to turn the positive diacetate into the negative
triacetate, at room temperatures. As cellulose triacetate fibers are
heated, the temperature for *zero* birefringence (T_{ZB}) is reached, above
which the fibers become optically positive. This phenomenon is explained on the basis of increasing diameter of the radially polarizing
acetyl group as the temperature increases; thus the dipole moment of
the acetyl group decreases to the point where the axial polarizability
takes over and the fiber becomes optically positive.[10]

Group and Bond Refractions

The foregoing remarks about the molecular basis for anisotropy in
fibers and other materials is strongly related to the old established sys-

tem of atomic and molecular refractions used for almost a century by organic chemists to confirm the composition and the supposed structure of organic compounds.

We have seen above that the fundamental basis of the refraction of light by a material rests upon the interaction of the radiation with the electrons of the material, and hence depends on the nature, the spacing, and the charge distribution within the atoms that control these electrons. The theoretical relationship between refraction and density derived in 1880 by Lorentz and Lorenz is

$$\text{molar refraction} = \frac{n^2 - 1}{n^2 + 2} \frac{M}{\rho}$$

where n is the refractive index of a pure substance at a given temperature and wavelength, ρ is the density at the same temperature, and M is the molecular weight. The dimensions turn out to be those of volume. Like the other forms of molar volume, the molar refraction is an additive and constitutive property,[22] and so was soon resolved into atomic refractions such as those for carbon, oxygen, sulfur, and so on.[22] A more useful extension of the principle makes use of *group* and *bond* refractions, such as those which have been established for organosilicon[23] and organotin[24] compounds. Here the measure of density and refractive index for hundreds of compounds leads to surprisingly constant values for bond refractions which can be relied upon to corroborate or dismiss a suggested pattern of alkyl and aryl bonds with C, N, O, Si, S, halogens, etc., in any newly synthesized compound.

Turning the concept of molar refraction around, we see that we may apply it to predicting the refractive index of *compounds* of known composition and structure. If we go a step further and think of the *oriented* portions of known compounds of very high molecular weight, whether oriented by internal forces during crystallization or by external forces during stretching or other deformation, we see that there can be different refractive indices in different directions within the macromolecular structure. For example, an oriented long chain of singly bonded carbon atoms with protruding side groups containing aryl groups, oxygen, nitrogen, sulfur, or halogens will have one predicted refractive index in the direction of the carbon backbone and a quite different predicted refractive index crossways to the backbone. The difference between the two indices constitutes molecular anisotropy. However, what happens to the molecular anisotropy in a fiber or other material depends on the kind and extent of crystallization, orientation, and stress in that particular fiber or other phase of material. The atomic refractions of heavy elements are much higher than those of light elements, as shown in

Glasstone's *Textbook of Physical Chemistry*.[22] For example, the atomic refraction of iodine is 14 ml, while that for chlorine is 6 ml and that for carbon is 2.4 ml. Similarly, the bond refraction for Sn—I is 17.95 ml, while that for Sn—O is 3.84 ml. In a practical way, the glassmaker knows this when he adds lead oxide or barium oxide to his glass composition to increase the refractive index of the glass.

The chemical composition of a fiber can be varied within the limits of nomenclature set by the U.S. Government.[18] Thus acrylic fibers must contain at least 85% acrylonitrile; fibers containing mostly acrylonitrile, but less than 85%, are termed modacrylic. In both types the additional fiber-making material can be added as monomer(s) to make copolymer(s) (practically one kind of molecule) or as polymer(s) to make a mixture of molecular species. Moreover, different kinds of molecules, coming from different but adjacent spinnerettes, may be spun together to produce different textures and other fiber properties. All of these manipulations of molecules vary the optical properties of the fibrous products.

Variations in molecular composition affect far more than the polarizability of the individual molecules. As we have already seen, the optical and mechanical properties of fibers are also affected by the *organizability* of molecules. They may be organized as single crystals, as spherulites or other multiple crystals, as mixed multiple phases (resulting in form-birefringence), or as stress-oriented molecules (as a result of stretching, crimping, or dual spinning).

6.4. ANISOTROPY: MOLECULAR ORIENTATION AND ORGANIZATION

Anisotropy due to the orientation and organization of molecules into fibers occurs in many steps of growth or manufacture. Here we shall confine ourselves to man-made fibers, since the steps are designed to vary mechanical properties and since much can be learned about the process from the various optical properties.

1. *Spinning.* The organization, orientation, and distribution of molecules begin with spinning, either spinning from the melt or spinning from an aqueous ("wet") solution or an organic ("dry") solvent. The spinnerettes are designed to vary the external shape of the fiber, its internal solidarity or hollowness (number of canals), and, by dual spinning, the texture of the fiber. The kind of liquid and the design of the spinnerette can affect the orientation of the molecules radially (skin-and-core structure) as well as axially (flow birefringence).[13,17]

2.   The method and rate of *coagulation* or jelling can affect the optical and mechanical properties of the spun product.[7] The process can be simulated to some extent under the microscope.[19] Postmortems can be performed on commercial lots of dry-spun vs. wet-spun fibers.[13] Dry-spun fibers can be coagulated quickly and wet-spun fibers can be coagulated slowly. Fast spinning can favor skin formation over slow spinning.[13]

Scott showed with experimental polyethylene terephthalate ribbons that drawing conditions and treatment can produce a variety of structures and corresponding properties.[17] The variation originates with the type of neck that is produced in the beginning of the stretch-reduction process (necking-down). Figure 6.11 gives the shapes of three types of necks. Figure 6.11a shows the usual shape with most plastics; it has sloping shoulders and a concave throat.[7] The core stretches before the skin, if any. Polyester, as spun, draws this way,[17] but if the spun polyester is first aged and then drawn, the skin draws before the core, producing the second type (Figure 6.11b), the lustrous kind. Figure 6.11c shows the third type, produced by drawing the aged as-spun polyester over a knife edge so that the side away from the edge draws first, producing a clear product of elongation.

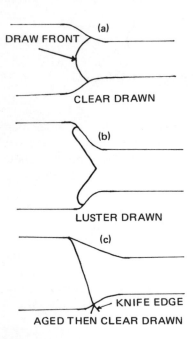

FIGURE 6.11.   Scott's three types of reduction of an experimental polyethylene terephthalate ribbon.[17]

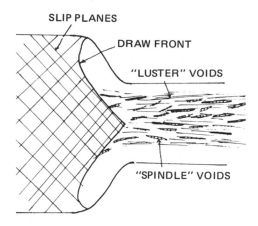

FIGURE 6.12. Another view of Figure 6.11b (Scott's diagram), polyethylene ribbon, showing slip planes, draw front, and voids.[17]

Figure 6.12 shows diagrammatically the slip planes and lines, the slip front and filmy luster voids. It also shows the spindle voids which are typical of polyester aged before drawing, whether or not it also has filmy, luster-producing voids. Both types of voids probably contribute to *form-birefringence*. The degree of contribution can be checked by measuring the birefringence immediately after mounting in a penetrating liquid and again after penetration of the liquid,[6] or, as Scott did,[17] by measuring birefringence in thick and thin sections.

The role of *aging* is to allow sufficient time (4 days) for crystallization to take place, to the extent of about 40%. At the end of this much time, Scott observed the presence of spherulitic nuclei. These nuclei tie molecules together, stiffening the as-spun structure to the extent that slip planes form during deformation, as in metals.

Thus we see that the origins of birefringence in fibers are several: inherent in molecular structure, developed by crystallization, produced by form-orientation of two or more phases, or developed by stress.

6.5. TRANSPARENT SHEETS, FOILS, AND FILMS

Strain as a result of stress in inorganic or organic glass is visible between crossed polars. Indeed, organic sheets such as those of poly(methyl methacrylate) are used to construct models of structures such as bridges and buildings so as to study the location and extent of strain under stress by means of the photoelastic effect seen between

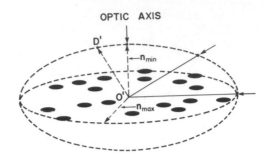

FIGURE 6.13. Planar molecules in sheets or foils.[16]

crossed polars. Some plastic sheets, such as those used as windows, are prestressed uniaxially or biaxially. Transparent foils and ribbons are often used to study dynamically the effects of tension or compression.[20,21]

Planar molecules are especially interesting if the molecular planes lie in the plane of sheets or foils, as shown in Figure 6.13, for then a uniaxial (or biaxial) interference figure is visible.[16]

6.6. SUMMARY

The use of the polarizing microscope for studying fibers brings all the accumulated techniques of optical crystallography and petrography to bear upon the identification and characterization of textile fibers, creating a powerful analytical science to help a rapidly developing technology. At the same time, it is not necessary to know about classical crystallography in order to study fibers. All that is required is to consider the fiber as a sort of simplified uniaxial organization and to apply the apparatus and techniques of polarized-light microscopy as developed in Chapter 5.

Fibers have at least eight determinative optical characteristics which identify them and are very useful in research. Not all fibers exhibit all eight, but the very lack of some serves as a distinguishing characteristic. To determine these properties, a few 1-cm-long portions of the fiber in question are mounted in oil (if hydrophobic) or in water (if hydrophilic), covered with a cover glass, brought into focus, and examined on the rotating stage of a polarizing microscope between crossed polars.

The first characteristic to look for is brightness in some or all positions of rotation, indicating anisotropy. Glass fibers remain dark because they are very thin and the glass is very nearly isotropic, being limited to a small amount of strain birefringence at most. Most man-made organic fibers are bright in certain positions of rotation because they are uniformly anisotropic, with their units primarily parallel to the fiber axis. If the anisotropy is weak or questionable, insert a first-order red plate so that the field will be red instead of black. If the fiber is anisotropic, it will change color upon rotation of the stage, being bluish in one set of diagonal quadrants and yellowish in the opposite set. Cotton appears bright in all positions because of its spiral structure.

The second point to look for is whether the fiber "winks out" (shows complete extinction) every 90° rotation. Cotton does not; linen departs more and more from complete extinction as it gets older, because its fibrils become unfastened from their bundles; silk shows short portions with different extinction due to crossing of the still plastic filaments in the cocoon. Man-made fibers almost always show complete extinction at some position, unless they have been crimped or otherwise forcibly deformed.

Third, the extinction may be completely or partially parallel or oblique to the fiber's long direction. Man-made fibers usually show parallel extinction because the structural units are arranged in parallel chains by the spinning and stretching processes. The microscopist must be sure that the directions of vibration of polarizer and analyzer are parallel to the cross lines in the ocular, of course.

Fourth, in anisotropic fibers the amount of retardation of one emerging wave train vs. the other is determinative. The retardation is manifest as grays, white, or colors between crossed polars, depending on the thickness and degree of anisotropy. The birefringent fiber splits the polarized beam into two polarized rays, both of which must go through the analyzer. There they interfere because of their phase difference, and interference colors result. These may be of several orders, of which the second order is brightest; after the fourth order the colors are faded and the blues and yellows disappear. The order of the color may be established by using a quartz wedge of several orders of interference colors, counting the orders as they go up or down as the wedge is slowly inserted and removed. Acrylics and cellulose acetates show first-order gray or color, whereas drawn polyester fibers show orders higher than three. The difference in order of colors between stretched and unstretched nylon, polypropylene, or polyesters is so great that degrees of stretching (drawing) can be recognized.

Fifth, the *sign* of retardation is important. If light moves more slowly when vibrating lengthwise instead of crosswise (refractive index of light vibrating along the fiber, $n_{||}$, greater than refractive index for vibrations across the fiber, $n_\perp$), then the fiber is said to be positive.

Most fibers are positive, but saran, vinyon, acrylics, and cellulose triacetate are negative. To determine the sign of a fiber that shows first-order gray, the fiber is oriented in the same direction as the slot for the first-order red retardation plate. As the plate is inserted, if the fiber is positive it will appear yellow; if negative, it will appear blue against the red background. The reasons for the change are explained earlier in the chapter. A quarter-wave plate or a quartz wedge may also be used to indicate the sign of birefringence, since they distinguish between addition or subtraction of the two retardations. Some colored fibers are dichroic (that is, they absorb more light or more of some colors in one direction than another), and since the greater absorption usually occurs in the direction of higher index, the dichroism often may be used to indicate the sign of birefringence. In positive fibers the greater absorption will occur in a direction parallel to the fiber, and in negative fibers it will be perpendicular.

Sixth, the *degree* of birefringence of a fiber is strongly determinative. The degree may be expressed as the difference between the two refractive indices $n_\parallel$ and $n_\perp$. This difference will be only 0.001–0.01 for weakly birefringent fibers such as proteins, acrylics, acetates, saran and vinyon, but stronger (0.01–0.1) in cellulose, vinyls, silk, nylons, and polypropylene, and strongest in polyesters. The difference can be estimated from the measured fiber thickness and the first-order gray or color it exhibits by using the Michel-Lévy chart. Fibers of greater birefringence display several color bands parallel to the axis because of their cylindrical (or other symmetrical) shape, and the band of highest order is used in consulting the Michel-Lévy chart. Thickness of a fiber may be measured by bottom-to-top focusing, or by looking at an end-view with a micrometer eyepiece.

Seventh and eighth, the *actual values* of refractive indices $n_\parallel$ and $n_\perp$ are characteristic properties subject to measurement. Refractive index is measured by immersing the fiber in successive standardized liquids of known refractive index until a match is obtained. Here the polarizer is used alone, without analyzer, and the fiber is lined-up first parallel to the direction of vibration in the polarizer and then perpendicular to it. If strongly axial illumination is used, the Becke line (refraction halo) will move *toward the medium of higher refractive index* as the objective is focused *upward;* with oblique illumination (half a cone of polarized light), a dark shadow will appear on the side of the fiber opposite the dark side of the field if the fiber is below the refractive index of the liquid. In either case, the liquid is changed repeatedly until a match is obtained, usually from a set of standard liquids. The *difference* between the measured $n_\parallel$ and $n_\perp$ gives the quantitative birefringence independently, and of course also the sign of the birefringence. The values of $n_\parallel$

and $n_\perp$ are obtained at the *surface* of the fiber which is in contact with the standard test liquid. However, if the fiber structure is *not radially homogeneous*, the surface value and sign of birefringence calculated from $n_\parallel$ and $n_\perp$ will differ from the *average* value and sign of birefringence (characteristics 5 and 6) obtained from the *total* retardation and thickness of the fiber by means of the Michel-Lévy chart.

The magnitudes of the refractive indices are related to the chemical composition and the polymeric structure of the fiber in some general ways that are discussed earlier in the chaper.

Microscopical Properties of Crystals

7.1. STRUCTURAL CLASSIFICATIONS

A solid *crystal* is composed of atoms, ions, or molecules arranged in a pattern which is periodic in three dimensions.[1] There are seven *structural*[1] classifications: *isometric* (cubic[1]), *tetragonal, hexagonal, trigonal* (rhombohedral[1]), *orthorhombic, monoclinic,* and *triclinic.* The trigonal classification (of which calcite, the rhombohedral phase of $CaCO_3$, is a typical example) can be considered to be in the hemihedral class of the hexagonal system, being made up of the alternate faces of a hexagonal bipyramid. *Structure* determines the inherent properties of a crystalline compound. For example, the three crystalline phases of $CaCO_3$, calcite, aragonite, and vaterite all have different crystalline structures, different solubilities, and different melting points, hardnesses, and optical properties.

7.2. MORPHOLOGY

Crystals are also characterized by their *morphology*,[1] that is, their sizes and shapes, as grown from solution, from melt, from vapor, or by alteration of other crystals. The morphology of a crystal determines its rate of change, such as its rate of dissolution, melting, or other phase transformations, because morphology controls its specific surface. Grown slowly from solution, a crystal manifests characteristic *faces* which are definitely related to the structure. Thus one kind of face, called a *form*,[1] has constant angles, whether the crystal be equally developed (equant), long (columnar or acicular), or flat (tabular or lamellar).[2] That is, for *each* of the six or seven kinds of structure there are *five*

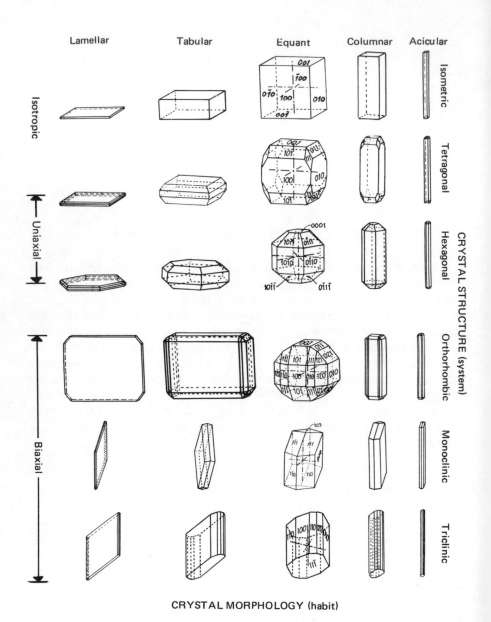

CRYSTAL MORPHOLOGY (habit)

FIGURE 7.1. Crystal structure and morphology.[2]

kinds of naturally polygonal morphology, called *habit*, as illustrated in Figure 7.1.[2]

Crystals grown from solution[3] or by precipitation[4,5] on a microscopical slide are seldom equant because the drop of mother liquor is shallow and is difficult to stir while under the objective. Common salt (NaCl), for example, recrystallizes from a drop of solution on a slide not so much as equant cubes but as flat and long rectangular shapes (Figure 7.1), sometimes tipped to make interpretation all the more challenging. The terms for habit have variations in the literature,[3-5] but the structure remains constant. For this reason, the face angles and optical properties are also constant, no matter what the habit.

7.3. MILLER INDICES

Figure 7.1 shows the *Miller indices* for typical faces shown on equant habits. These indices are the reciprocals of the fractional intercepts that a face makes on the crystallographic axes. The axes are graphic (typical directions of spacing of the structural units, be they atoms, ions, or molecules). The form representing the isometric system in Figure 7.1 is the cube, a special hexahedron in which all of the face angles are right angles. It is a closed form, that is, all of the faces are structurally alike (regardless of the dimensions of the edges in the flat and long habits). As the axes are set up in the equant habit in Figure 7.1, the front face cuts one axis by one unit (1) and the other two axes at infinity (∞), so the Miller indices are $1/1:1/\infty:1/\infty = 100$. The back face is the same, but to show that it is cut in back by one unit, its specific Miller index is $\{\overline{1}00\}$. The right side is written $\{010\}$ and the left $\{0\overline{1}0\}$; the top $\{001\}$, and the bottom $\{00\overline{1}\}$. The typical index for the cube is simply written $\{100\}$. We could, of course, set up the axes diagonally and then the typical index would be $\{110\}$, but in this instance nothing is to be gained by it. The Miller indices of the lamellar, tabular, columnar, and acicular habits are exactly the same as those of the equant habit (cube). It does not matter that the edges are of various lengths, except that there are more unit cells (all isometric) attached along the longer edges. How do we prove isometry microscopically? If the crystals are isotropic (black) in every orientation as observed between crossed polars, they demonstrate that they are isometric.

It is easy to see why a cubic crystal is isotropic. Since the unit spacing is the same along the three axes, the velocity of light (and refractive index) for a particular wavelength is the same in any direction at any one temperature. For sodium chloride, NaCl, the refractive index, $n_D^{20°C}$, is 1.544. For potassium chloride, KCl, $n_D^{20°C}$ is 1.490. While NaCl

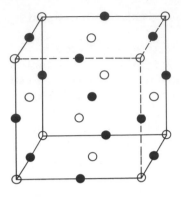

FIGURE 7.2. Isometric ionic lattice. ● represents positive ion, such as K^+ and NH_4^+, interchangeably. ○ represents negative ion, such as Cl^-, Br^-, I^- (interchangeably).

and KCl are both isometric and isotropic, they are *not isomorphous;*[1] that is, they do not crystallize together as mixed crystals in *solid* solution. Sodium and potassium ions are not interchangeable in the crystal structure (Figure 7.2), but ammonium ions are interchangeable with potassium ions in crystals of their chlorides to make a continuous isomorphous[1] series with refractive indices varying from $n_D^{20°C} = 1.49$ for pure KCl to 1.64 for pure NH_4Cl, in direct proportion to the potassium and ammonium content.[6] Hence the microscopical refractive index of a mixed crystal, coupled with the isotropy, is very descriptive and determinative, once the presence of NH_4^+ and K^+ is established.[4] Rubidium and cesium ions are also interchangeable with K^+ and NH_4^+, and so could influence the optical properties, but their absence of presence can be detected by other tests.[4]

If NaCl in solution contains a small amount of OH^- ion, such as from urea, *octahedral* faces appear on its crystals, with or without cubic faces.[7] The Miller indices of the octahedral faces of NaCl are all {111}; that is, all three axes are intercepted equally. The structure of the NaCl remains the same: isometric. Therefore, octahedral NaCl remains isotropic, with the same refractive index, $n_D^{20°C} = 1.544$, as cubic NaCl.

7.4. ISOMORPHISM

The alums are a remarkable set of isometric-isotropic crystals with interchangeable ions, all patterned after their namesake, potassium aluminum sulfate dodecahydrate, $K_2SO_4 \cdot Al_2(SO_4)_3 \cdot 24H_2O$). The *water of hydration* of an alum, too, is in the space lattice[8] because it is in the coordination sphere of the ions. The K^+ may be replaced by NH_4^+, Na^+, Cs^+, Rb^+, or Ag^+. The Al^{3+} may be replaced by Cr^{3+}, Fe^{3+}, Mn^{3+}, In^{3+}, Ga^{3+}, or Tl^{3+}; the SO_4^{2-} may be replaced by SeO_4^{2-}.[3] Moreover, a crystal of one pure alum (e.g., a colored one) will act as a foundation for an overgrowth of another (e.g., a colorless one). The substitution of various ions one for another, and the overgrowth of one alum on another, are both manifestations of isomorphism.[3] All the interchangeable ions fit into specific places in a single crystalline pattern that is *isometric* in structure and *octahedral* in habit. True, the refractive index varies with composition, but it varies directly with the proportional content of substituted ion. This is an analytical advantage, for it allows rapid determination of composition once the constituent ions are known. Some other microscopical tests often identify the ions, too: Cesium alum, for example, is much less soluble than the other alums and can be detected as it precipitates at low concentration from a drop of solution on a slide.[4]

7.5. SKELETAL MORPHOLOGY

The morphology illustrated in Figure 7.1 is of mature euhedral[9] crystals; no matter how small, they are filled out to the full extent shown. Immature crystals can look different and be misinterpreted. If, for example, crystals of NaCl are grown from a very shallow drop (one that has spread over too much of the microscopical slide), *hopper-shaped* crystals may develop. Because the drop is so shallow, the top face has run out of solute, while the meniscus of mother liquor attached to each vertical wall has supplied solute to build the walls higher than the top surface. As such a crystal grows, it becomes dished like a rectangular hopper. Unless one focuses up and down with an objective of short depth of focus (high NA), or uses oblique illumination, he may misinterpret the depression for an elevation—indeed, for a set of faces.

Dendrites are even less filled-out crystals. They are fernlike, treelike, branchlike skeletal crystals which have been grown so fast that they have not had the time to form faces and therefore are *anhedral*.[9]

Crystals which have grown from the melt[3,10,11] usually meet their neighbors in more or less straight, polygonal boundaries of crystalline *grains*, as in igneous rocks, cast metals, frozen water, and other solidified liquids. This action may be observed by watching water or other liquids freeze on a cold stage.[4] Many organic and some inorganic solids may also be observed during melting and freezing on a hot stage.[3,10,11] Thus the dynamic processes can include changes in temperature gradients, annealing and quenching, concentration and distribution of components and impurities, changes in morphology (sizes and shapes of grains), and variations in structure (polymorphic transformations). Such knowledge can lead to better understanding of materials as they undergo changes during manufacture, bulk-testing, usage, storage, weathering, or failure. The experience is also very helpful in forensic investigations.[12] Usually metals, ceramics, and other inorganic systems melt at too high a temperature to be studied directly on the hot stage of a microscope. In such cases the experiments are conducted elsewhere, and samples are taken and studied "postmortem" under the microscope. Here the experience with low-melting-point materials as they undergo dynamic changes will aid in interpreting the appearances of high-melting-point or opaque materials.[10]

The polyhedral boundaries of crystalline grains, such as those grown from the melt in igneous rocks and certain ceramics, are the result of close packing, like crowded soap bubbles. A crystalline grain can be a single crystal, a twinned crystal, or a spherulite. Each has its characteristic optical properties. Transparent single crystals grown from the melt can be studied microscopically like euhedral[9] ones, except that the optical properties usually cannot be related to grain boundaries because they do not consistently represent crystallographic faces. Yet there are certain optical properties of anhedral[9] crystalline grains that can be interpreted in terms of the crystallographic axes, and these are worth noting. For example (Figure 7.1), in the *tetragonal* and *hexagonal* (including trigonal) systems there is only *one unique axis.* Such crystals are *uniaxial* because the unique axis represents the only direction down which the perpendicular field of view appears to be isotropic. That is, viewed along that axis the field is dark throughout 360° of rotation between crossed polars. This occurs because the unique axis (the c-axis) is at right angles to the plane of the other two (tetragonal, a, a) or three (hexagonal, a, a, a) along which the spacing is the same; so in looking down the c-axis, the isometric plane likewise is perpendicular to the axis of the microscope, and we see the *one plane of isotropy.*

In the *orthorhombic, monoclinic,* and *triclinic* systems (Figure 7.1) the spacings are different in all three crystallographic directions, so there is no plane of equal spacing. However, there are two directions in each of which the field of view is isotropic (dark throughout 360° of rotation of the specimen between crossed polars). These, the two *optic axes,* are the directions in which light travels with no apparent birefringence.[3] A bisectorial view includes both optical axes *if* the angular apertures of the condenser and the objective are wide enough. It follows that there is an *acute bisectrix*[1] and an *obtuse bisectrix,*[1] and each is recognized by its characteristic interference figure (see also Chapter 5).

These, and some other crystallographic and optical properties, may be determined on unicrystalline grains in thin sections (≈ 30 μm) of rocks, minerals, and man-made products[6,13] within a wide range of grain sizes. Searching for grains of adequate orientation[14-16] may be time-consuming. If so, a universal stage[17] to orient the specimen conveniently may be desirable.

TABLE 7.1
Distribution of 935 Crystalline Phases among
Isotropic, Uniaxial, and Biaxial Crystals[6,18]

System	Number of phases		% Total, inorganic plus organic
	Inorganic[a]	Organic[b]	
Isotropic	86 (17.0, 9.2)	7 (1.6, 0.8)	10.0
Uniaxial	101 (19.9, 10.8)	49 (11.5, 5.2)	16.0
Biaxial	315 (62.1, 33.7)	359 (83.9, 38.4)	72.1
Intermediate	5 (1.0, 0.5)	13 (3.0, 1.4)	1.9
Totals	507 (100.0, 54.2)	428 (100.0, 45.8)	100.0
Grand total,[3] number of phases, 935			

[a] The values in parentheses represent, respectively, the % of the total number of inorganic phases, and the % of the total number of inorganic plus organic phases.
[b] The values in parentheses represent, respectively, the % of the total number of organic phases, and the % of the total number of inorganic plus organic phases.

If it can be determined whether a crystal (euhedral or anhedral) is isometric, uniaxial, or biaxial, a classification of sorts can be made on this basis alone. Thus, Table 7.1 shows the distribution of 935 frequently encountered crystalline phases, data compiled by Winchell[6,13] and classified by Kirkpatrick.[18] Only 10% of the 935 are isotropic, and most of these are inorganic (probably because the coordination spheres of the constituent ions require such a high degree of symmetry). Sixteen percent of the 935 phases crystallize in the uniaxial systems, one-third of them organic compounds. Seventy-two and one-tenth percent of the 935 phases are biaxial, more than half of them organic compounds, showing how the relatively large molecular units of the organic compounds must pattern themselves.

7.6. ISOTROPIC SYSTEM

For a given wavelength and a given temperature, there is only one value for the refractive index of an isometric crystal, since it is isotropic. Figure 7.3 is a chart[18] showing the distribution of the approximate refractive indices (with white light at room temperatures) for inorganic isotropic substances. The peak value is about 1.45 (25% of total), for values falling between 1.33 and 2.0.[3] For each species of isometric crystal there is some variation in refractive index with respect to the wavelength, called *optical dispersion*.[1] For example, the refractive index of $2Na_3PO_4 \cdot NaF \cdot 19H_2O$ for the sodium D line is 1.452, and the refractive index for the sodium C line is 1.451. The optical dispersion is

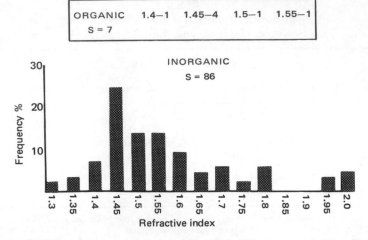

| ORGANIC | 1.4—1 | 1.45—4 | 1.5—1 | 1.55—1 |
| S = 7 | | | | |

FIGURE 7.3. Distribution of the approximate refractive indices (with white light at room temperature) for inorganic isotropic substances.

$n_C^{25°C} - n_D^{25°C} = 1.451 - 1.452 = -0.001$.[19] Dispersion is just as much a determinative optical property as any other. The degree of dispersion is important also in the technique of *dispersion staining*,[20,21] in which differences in refractive index for some wavelengths (but not others) between crystals and mounting liquid will lead to colored outlines (Chapter 9).

It is relatively easy to recognize octahedra {111} when they are equant, for they have the appearance shown in Figure 7.4a. The octahedron is shown lying on one of its equilateral triangles (dotted). The

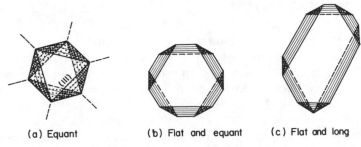

(a) Equant   (b) Flat and equant   (c) Flat and long

FIGURE 7.4. Drawings[3] of octahedra, each lying on an octahedral face and all three showing the same angular symmetry.

opposite one is shown in Figure 7.4a (solid outline), pointed the opposite way. The other six sides are *alternately* pointed *up* or *down*, connected by edges that alternately go from top to bottom or vice versa. These six zigzag edges, *in perspective*, make an equilateral hexagon. It is not to be confused with a truncated, bipyramid in the hexagonal system (or anything else), since there are the *doubly shadowed* up *and* down portions which look like smaller, darker triangles in perspective. These small dark triangles are also shown in the *flat* octahedra (Figures 7.4b and c), but the equant and flat octahedron (Figure 7.4b) shows in perspective 12 side lines instead of 6, as shown in Figure 7.4a. The long and flat octahedron (Figure 7.4c) is most difficult to interpret because the side lines are not of equal length. However, the *angular* symmetry is still the same and that is what matters as an indication of the *isometry*, and the isotropy clinches the case. Incidentally, the shadows (shading) in all three drawings (Figure 7.4) represent the partial reflection of incident light away from the microscopical objective. The degree of shading represents the obliquity of the side faces.[3]

Besides the cube and octahedron mentioned above, the isometric system is represented (less frequently) by the rhombic dodecahedron (as of hexamethylene tetramine) and pentagonal dodecahedron (as of cerium formate). There is also the hemihedral class of the octahedron, i.e., the tetrahedron (as of sodium uranyl acetate).[4]

7.7. UNIAXIAL CRYSTALS

In the *tetragonal* and *hexagonal* systems (Figure 7.1) there is one unique axis, namely, the c-axis, along which spacing of the structural units is shorter or longer than along the a-axes which lie in a plane perpendicular to the c-axis. In the tetragonal system there are two a-axes at 90° to each other; in the hexagonal system there are three a-axes at 120° to each other. In either case, in any plane, light vibrates in one direction at a constant velocity, and therefore the ordinary refractive index ω is always "seen." On the other hand, the extraordinary refractive index ϵ is "seen" only if the crystal is lying on a prism face such as {100} or {110}. On an inclined face (e.g., pyramidal or rhombohedral) the refractive index for light vibrating at right angles to the always "seen" ω is somewhere between the values for ω and ϵ, and is represented by the symbol ϵ'. For crystals lying on a definite face inclined to the c-axis (such as a pyramidal or rhombohedral face), the value for ϵ' is *constant* for that *particular form* (kind of face). Values for ϵ' are not generally given in compilations,[2,6,13] so the reader would do well to start his own collection of ϵ' values in some retrievable manner. It is emphasized that the values ω, ϵ, and ϵ' are for vibrations which are crystallographically oriented. If there is any doubt as to the orientation of the axes, the more abstract and general symbols,[9] such as n_1 and n_2 should be used.

The plane of the a-axes is the basal pinacoid. Crystals resting on this face appear dark in every position of retardation between crossed polars.

Most uniaxial crystals recrystallized on a microscopical slide do not generally rest on a basal pinacoid, but some species do (for example, iodoform, lead iodide, cadmium iodide, and Na_2SiF_6). Uniaxial crystals which have low-enough melting points, such as sodium nitrate, may be melted on a slide, covered with a cover glass, and recrystallized from the melt. The resultant crystalline *grains* may be searched for the isotropic view. If either of these two methods fails to produce an isotropic view, and if a universal stage[17] is available, it may be employed to rotate the crystal 180° in three perpendicular directions. Such a stage may be used with almost any small single crystal, be it a polygon, fragment, sand grain, sawed section, or product of fusion.

Once an isotropic view is sighted, the top lens of the condenser is put in place and the diaphragm is opened all the way to give *conoscopic* illumination and an interference figure.[3] (An objective of NA = 0.85 is preferred if its presence is assumed in tables of computation.[14]) Next the Bertrand lens (Chapter 5) is put into place in the microscope's tube, below the ocular, to provide a compound microscope focused on the *back aperture* of the first objective. If no Bertrand lens is provided, the ocular is removed and the back aperture of the objective is viewed directly, i.e., without magnification. If the crystal is uniaxial, and if the microscope is properly aligned, the isotropic view provides a *centered uniaxial interference figure* or pattern. As shown in Figure 7.5, there is a dark Maltese cross ringed with interference colors of numerous orders, depending on the degree of birefringence. The rings of interference colors of Newton's series and Michel-Lévy's chart represent the increased retardation derived from increased angularity from the axis to the periphery of the cone of illuminating rays. In other words, the rings of *isochromes* are summations of retardations corresponding to the variations in orientation of incident rays, with reference to the c-axis of the crystal. The black arms of the Maltese cross are called *isogyres* of the interference figure.[3] The cross is similar to the one shown by spherulites, such as starch grains, and for the same reason. They represent positions of extinction $\updownarrow$ (N/S) and $W \leftrightarrow E$. Instead of a sphere of crystallites, we have a sphere of directions. Indeed, we may get the same effect by illuminating a sphere made of the uniaxial crystal with axial light, employing crossed polars.

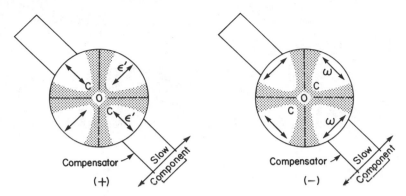

FIGURE 7.5. Diagram of uniaxial interference figures, positive (left) and negative (right). Directions of vibration of slower components are indicated by arrows. c, c represent quadrants in which compensation occurs. o represents optic axis.[3]

What good is an interference figure? If it indicates that the crystal is *uniaxial*, we know immediately that we are dealing with a species which represents only 16% of a total of 935 crystalline species (Table 7.1). If the substance is *organic*, it is among only 5%; if it is *inorganic*, the probability is 11%. Thus, uniaxiality alone is distinctive and determinative.

Moreover, the optic sign is easily obtained from an interference figure.[3,15] If a first order (1°) red retardation plate is inserted into the NW\SE diagonal with its slower component vibrating SW⁄NE, the color near the center of the (uniaxial) black cross is yellow (subtraction) in the NW and SE quadrants, and the crystal is optically positive. If, instead, the color is blue (addition), the crystal is negative. Or, the quarter-wave ($\frac{1}{4}\lambda$) plate may be used. Then if a black dot is seen NW and another is seen SE of the center of the black cross, the crystal is positive; if NE and SW, it is negative. Or else, a quartz wedge may be moved in place with increasing order of interference color. If the crystal's colors move *outward*, it is positive; if *inward*, it is negative.[15] From Figures 7.6 and 7.7 we get a distribution (% probability), along the ordinates, for uniaxial positive and negative crystals among *inorganic* and *organic* species with regard to ω (abscissa), which is the sole value "seen" down the unique axis (c-axis). Thus, the combination of uniaxiality and quantitative value for ϵ' is very descriptive. Placed on punched cards or fed into a computer, the combined data become very selective, especially at the extremes of the distributions.

Off-centered figures may be used not only to obtain the optic sign but also to show the direction of vibration for ϵ'.[3] For a given path length, the more rings showing in the off-centered figure the closer ϵ' approaches ϵ in value. Therefore the measurement of the angle of tilt of the optic axis (c-axis), together with the value for ω, furnishes a way for

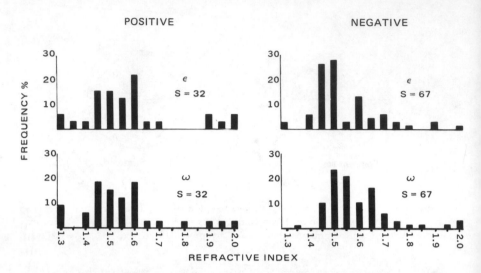

FIGURE 7.6. Refractive index distribution for uniaxial positive and negative crystals among inorganic species with regard to ω and ε.

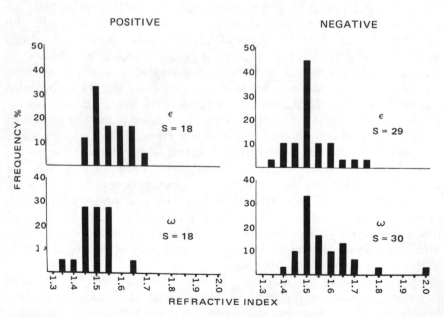

FIGURE 7.7. Refractive index distribution for uniaxial positive and negative crystals among organic species with regard to ω and ε.

the estimation of ϵ.[15] Determination of the differing values for ω and ϵ for different wavelengths of light establishes the degree of *dispersion*. In very distinctive rare cases, the optic sign may even change.

A *euhedral crystal* resting on a prismatic face {100}, {110}, {1010}, or {0100} presents ω and pure ϵ. These may be measured directly by any of the methods mentioned in Chapter 6, or given in other books.[3,15] The difference gives the birefringence and optic sign. If $\epsilon > \omega$, the crystal has a positive birefringence; if $\epsilon < \omega$, a *negative* birefringence. Alternatively, the birefringence and optic sign may be determined directly, as mentioned earlier. Moreover, a prismatic view presents best the phenomena of dispersion and pleochroism.*

A pyramidal, tetrahedral, or rhombohedral face presents ω, of course, but also a special and constant value for ϵ', which is therefore descriptive and determinative. Furthermore, if the face angle is known, data depending on true ϵ may be determined. Traces of the uniaxial interference figure may also be visible.

It is now obvious that attempts should be made to obtain crystal faces for observation. While crystals as commercially received are not always good enough for crystallographic and optical purposes, recrystallization from a drop of solvent on a microscopical slide[3,4] is generally fruitful.

7.8. BIAXIAL CRYSTALS

Table 7.1 shows that about 62% of inorganic crystals are biaxial, and even more organic crystals, 84%, are in this category. Figures 7.8 and 7.9 show distributions according to optic sign and the *three* principal refractive indices n_x, n_y, and n_z.[18] Since there are two *optic axes*, the angle between them $(2V)$ is also descriptive and determinative (Figure 7.10). This value is measurable or estimatable directly from the adequately centered biaxial figures (Figures 7.11a and b), whether the bisectrix is acute or obtuse. Measurement is made easier by standardizing on a numerical aperture of 0.85.[14] Then a standard chart for $2V$ vs. δ, the angle of rotation of the stage necessary to cause the isogyres (curves around the optic axes) to move from the center to the edge of the microscopical field, may be used to convert from $2E$ (optic axial angle in air).[15]

---

* *Pleochroism* is the general term for absorption of different colors as light passes in different directions through a crystal. If the crystal exhibits two distinctive colors when viewed in two unique directions in white light, it is said to be *dichroic* (see Chapter 6). If there are three directions and three colors, it is *trichroic*.

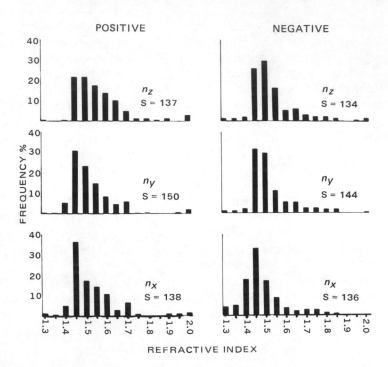

FIGURE 7.8. Refractive index distribution for biaxial positive and negative crystals among inorganic species.

It is probably apparent already that biaxial crystals are optically and crystallographically complicated (Figure 7.12).[22] This means that identifying the three principal refractive indices qualitatively and quantitatively may be difficult. But some values are always visible and capable of measurement, however vague their specific identity may be at the time of observation. The important thing is to record them and to recognize them on every repeated occasion, so that a pattern develops.[23]

7.9. OPTICAL PROPERTIES OF THE LIQUID-CRYSTALLINE OR MESOMORPHIC STATE

Some solid organic compounds assume the *mesomorphic* state, which is intermediate between that of a solid crystal and a true liquid. The mesomorphism which sometimes occurs when a solid crystal is heated is termed *thermotropism*; that which occurs when a solvent is added is called *lyotropism*.[24]

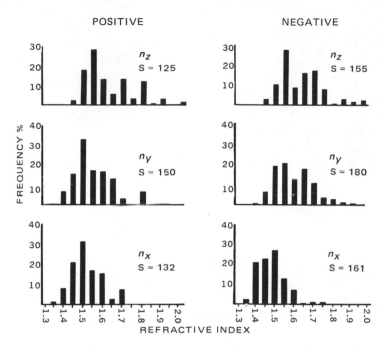

FIGURE 7.9. Refractive index distribution for biaxial positive and negative crystals among organic species.

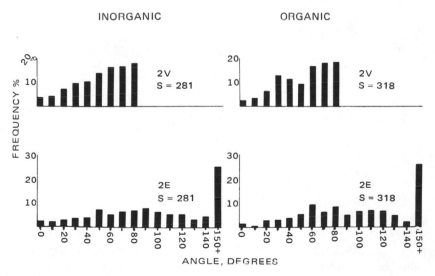

FIGURE 7.10. Optic axial angle distribution for biaxial positive and negative crystals among inorganic and organic species.

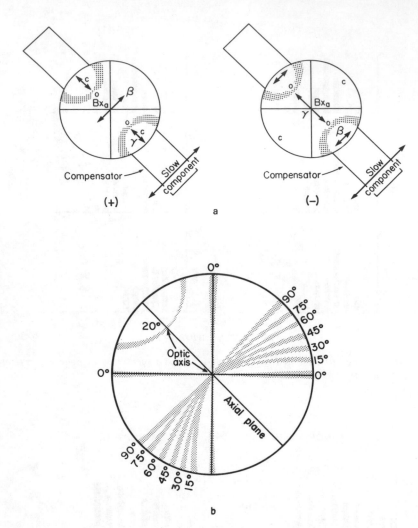

FIGURE 7.11. (a) Positive and negative biaxial interference figures. Directions of vibration of slower components indicated by arrows. *c, c*, quadrants in which compensation occurs. *o, o*, optic axes. *Bx_a*, acute bisectrix.[3] (b) Estimation of $2V$ from curvature of isogyre, in optic axis interference figure. (After F. E. Wright).[3]

Most compounds which manifest mesotropism have long molecules, sometimes also flat, such as those containing *para*-substituted benzene rings. Mesomorphic substances also contain one or more polar groups. In the solid crystalline state, the long and/or flat molecules arrange themselves in parallel positions, holding themselves together by means of the polar groups and van der Waals forces. When such crystals

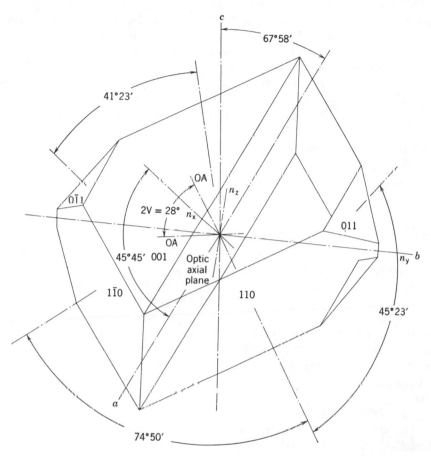

FIGURE 7.12.   Crystallographic properties of melamine.[22]

System	monoclinic	Refractive indexes	$n_x = 1.487$
Class	prismatic		$n_y = 1.846$
Axial elements	$a:b:c = 1.4121:1:0.9728B$		$n_z = 1.879$
	$= 112°2'$	Optic sign	$(-)$
Birefringence	0.392	Optic axial angles	$2E = 53°30'$
			$2V = 28°38'$

are heated or loosened with solvent, if the polar forces are not too strong, the molecules may find some freedom of movement while retaining some degree of order.[24]

Mesophases are birefringent. In lyotropic systems, form-birefringence (Chapter 6), having to do with difference in refractive index between liquid solvent and the long or flat solute molecules, may account for some of the birefringence, but not all of it.[24] The fundamental cause of birefringence in mesophases lies in the action of polarized light waves as they displace the electrons in the atoms toward the direction of vibration of the polarized light, producing electric dipoles oriented in the direction of polarization. When the direction of vibration is parallel (→) to the lineup of the dipoles (+ –)(+ –)(+ –), the total electric field will be *greater* than when the dipoles were widely separated. When the light is vibrating at right angles to dipoles which are side by side $\binom{+\ -}{+\ -}$, the total electric field is *less* than that for widely scattered atoms. So with long molecules, when light is vibrating parallel to the long axis, the average strength of the dipoles is greater (refractive index higher) than when light is vibrating crosswise.[24]

7.10. THERMOTROPIC, MESOMORPHIC, SINGLE COMPOUNDS

Generally, thermotropic mesophases of single compounds having long molecules belong to one of the following three types of structure:

1. *Smectic* mesophases is the name of the type to which soaps belong, as the Greek word for soap is *smega*. Soaps have a polar group such as —COONa at the end of a long hydrocarbon chain. The polar group of one long molecule attracts that of another molecule head-to-head, forming a double layer in the smectic phase, as illustrated in Figure 7.13. The layers are flexible and glide over one another. Most smectic phases are optically *positive* and uniaxial.[24]

2. *Nematic* mesophases are named after the Greek word for thread, *nema*, because some members of this type manifest thready lines. All members of this type are composed of single layers of more or less parallel lengthy molecules, as shown in Figure 7.14. Nematic phases are optically *positive* and uniaxial.[24]

3. *Cholesteric* mesophases are named after *cholesterol* and its derivatives because they are in the majority of members of this type. Theirs is a spiral arrangement of nematic layers as shown in Figure 7.15. The pitch of the spiral is about half that of a wavelength of visible light, giving the cholesteric structure very special optical properties including strong rotation of the plane of polarized light around the axis of the spiral (which is also the single optic axis). Cholesteric mesophases are optically *negative* because the long strands of such molecules are oriented in various directions around the axis. Since such molecules are also flat and spaced close together, the optical properties resemble those of planar molecules. When the layers are parallel to the slide, they may show an off-center biaxial interference figure with an optic axial angle of about 20°.[24]

OPTIC   AXIS

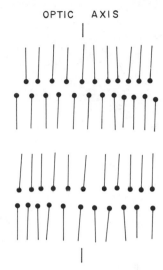

FIGURE 7.13.   Smectic structures (schematic): Double layers in soaps. (● represents terminal polar groups, e.g., —COONa.)[24] Courtesy of Microscope Publications, Ltd.

Some smectic types show polymorphism, believed to be due primarily to different lateral arrangement of the molecules. A smectic phase or phases may also manifest either a nematic or a cholesteric phase, but not both. The smectic phase always occurs at lower temperature than the nematic or cholesteric phase, since the smectic phase is more ordered. Because the crystalline phase of the same compound is still more ordered than its smectic phase, the crystalline phase is stable at a lower temperature.[24]

OPTIC AXIS

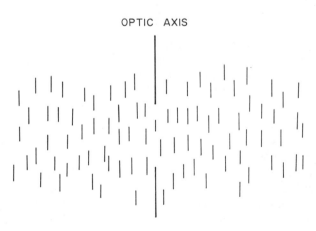

FIGURE 7.14.   Nematic structure (schematic).[24] Courtesy of Microscope Publications, Ltd.

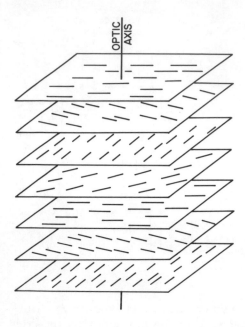

FIGURE 7.15. Cholesteric structure (schematic), with right-hand twist. The planes mark levels in the structure between which a 45° rotation of the molecular axes occurs.[24] Courtesy of Microscope Publications, Ltd.

7.11. KINDS OF MORPHOLOGY (TEXTURE)

The *morphology* of mesophases is called *texture* by Hartshorne.[24] There are several kinds of texture:

7.11.1. Homeotropic Textures

Homeotropic textures were named by Lehman and Friedel to designate smectic and nematic mesophases which have the optical axis perpendicular to the glass surfaces confining them. Hartshorne extends the meaning to include also cholesteric phases.[24] If, for example, a little (smectic) ammonium oleate is pressed into a thick layer between a glass slide and cover glass, a confused texture is seen between crossed polars because the specimen is too viscous for the layers to flatten out. If the glass surfaces are cleaned with hot soapy water without any etching agent, then rinsed with hot water, dried with a clean cloth and not handled thereafter, and if the cover glass is gently moved round and round by means of a rubber-tipped pencil, inspection between crossed polars reveals a gradual disappearance of the polarization interference colors. When the field becomes quite dark and the illumination is conoscopic, a uniaxial interference figure appears, as shown in Figure 7.16. By means of a first-order red plate, the optical sign is shown to be

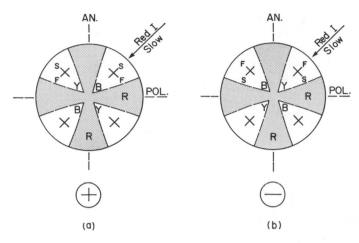

FIGURE 7.16. Determination of optical sign on optic axial interference figure by using Red I plate. (B = blue; Y = yellow; R = red.)[24] Courtesy of Microscope Publications, Ltd.

positive in the case of ammonium oleate because it is typically smectic. Typical nematic mesophases are also optically positive, but cholesteric mesophases are generally negative.[24]

7.11.2. Focal Conic Textures

Focal conic textures arise when a mesophase is confined to surfaces with which it forms strong local attachments, for example, to a glass slide and cover glass which have been etched with hydrofluoric acid. Around each center of attachment the long molecules are required to adopt a radiating arrangement that looks like a tipped hollow ring, known as a Dupin cyclide, as shown in Figure 7.17. The Dupin cyclide is truncated by the glass slide and cover glass. If the preparation is fairly thick and not too viscous, and is heated between a slide and cover glass (both slightly etched) until the liquid phase begins to appear, upon lowering the temperature and agitating the cover glass, a number of *polygonal* areas appear. In each polygonal area there is a family of ellipses consisting of large ellipses with smaller ones in the interstices. The major principal axes of the ellipses in any one polygon *meet in one point* within the polygon, as shown in Figure 7.18. If the upper surface of the preparation is in focus and the analyzer is in place (without the polarizer), *each ellipse is crossed by a straight isogyre, the narrowest point of which coincides with one of the foci of the ellipse.* If the *polarizer* is employed alone, a similar pattern will be in focus on the lower surface of

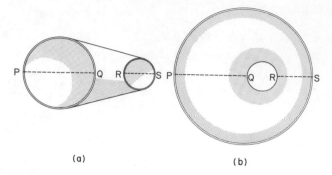

(a)   (b)

FIGURE 7.17. Dupin cyclide. (a) Half-cyclide showing principal sections, (b) plan view.[24] (After Hartshorne and *Stuart, Crystals and the Polarising Microscope,* Arnold, 1970.) Courtesy of Microscope Publications, Ltd.

the preparation. If the focus is gradually lowered from the upper surface to the lower surface of the preparation, it will be seen that each ellipse is joined by one branch of a hyperbola which starts at the isogyre. All of the hyperbolas belonging to the ellipses of one polygon meet in the lower surface at one point below the intersection of the major axes of the ellipses. This point is a common corner in a system of polygonal areas in the lower surface. Explanation of these phenomena is given by Hartshorne.[24]

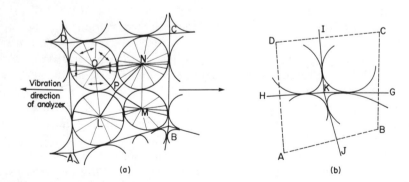

(a)   (b)

FIGURE 7.18. Polygonal texture. (a) Upper surface focused (analyzer, no polarizer), (b) lower surface focused.[24] (After Hartshorne and Stuart, *Crystals and the Polarising Microscope,* Arnold, 1970.) Courtesy of Microscope Publications, Ltd.

7.11.3. Other Smectic Textures

Other textures of smectic substances are:

1.   Bâtonnets, which are separate little images of focal conic textures, often highly ornamented. They may form on cooling the melt.

2.   Fanlike textures, which are revealed by crossing the polars, successive radial bands extinguishing at slightly different positions of orientation of the stage.

3.   "Oily" streaks, which may be observed between crossed polars when a focal conic texture is destroyed by shifting the cover glass on a preparation. The "streaks" are birefringent *bands* having *transverse striations*. At sufficient magnification the streaks' focal conic groups can be seen with the hyperbolas parallel to the striae. The bands are length-fast $(n_\| < n_\perp)$.

7.11.4. Nematic Textures

The *texture* of *nematic* mesophases is typically threadlike, formed by rapid cooling of the isotropic melt to the nematic temperatures. These wormlike textures mark structural discontinuities, analogous to ellipses and hyperbolas in smectic phases. However, the nematic discontinuities do not conform to any definite geometrical plan because there is no stratification. Approximately centered optic axis figures (Figure 7.16) or optic normal figures (Figure 7.19) are usually found with which to determine the optical sign (+). Incidentally, the initial morphology of the nematic phase from the melt is that of spherical droplets, not bâtonnets as from a smectic phase.[24]

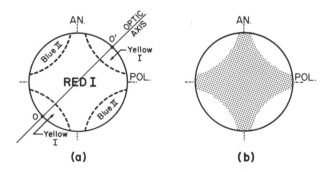

(a)   **(b)**

FIGURE 7.19.   Optic normal figure ("flash" figure). (a) 45° position showing example of color distribution. In a positive mesophase, light vibrating along O-O' (direction of the optic axis) is "slow." In a negative mesophase it is "fast." (b) Extinction position.[24] Courtesy of Microscope Publications, Ltd.

7.11.5. Cholesteric Textures

The *texture of a cholesteric* phase, produced by cooling the isotropic melt, is that of a cloud of very fine particles. When the cloud is disturbed by moving the cover glass, a homeotropic texture is produced, with the optic axis normal to the plane of the slide. The *vivid colors* are very different from those of homeotropic smectic and nematic phases. The specific color (or, occasionally, invisible wavelengths) depends upon the temperature and the angle of observation. The changes in color with temperature have been developed commercially to register temperatures and temperature gradients.[24]

The second characteristic of cholesteric phases is the *rotation* of the plane of polarized light. This phenomenon is manifested qualitatively under the microscope by lack of extinction between crossed polars. Instead of darkness there is a color which changes as the analyzer is rotated, due to the disperson of optical rotation. The optical activity of most cholestive substances is so great that *several turns* of the *rotating stage* are necessary to restore extinction. Therefore a circularly wedged preparation is used (Figure 7.20). The number of rings indicates the orders of 180° rotation by the specimen. A slight rotation of the analyzer to the right, noticing whether the rings go in or out, will indicate whether the specimen is levo- or dextrorotatory. A very low-power objective (≈48 nm) and eyepiece should be used. Since most cholesteric phases of single compounds are stable above room temperature, a hot stage should be used and the temperature of the test should be recorded as a factor in the degree of rotation by the specimen. Of course, a plane wedge may be made from a large cover glass propped up at one edge by another piece of cover glass; the specimen is run in between the sloping cover and a slide.[24]

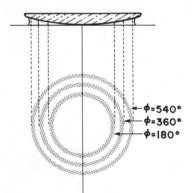

$\phi=540°$
$\phi=360°$
$\phi=180°$

FIGURE 7.20. Principle of determining optical rotation of a cholesteric phase.[24] Courtesy of Microscope Publications, Ltd.

7.12.   LYOTROPIC PHASES

Solutions of mesomorphic phases behave like cholesteric thermotropic phases.[24] Aerosol® OT, a substance with peglike molecules of sodium-di-2-ethylhexyl sulfosuccinate (a waxy smectic phase, difficult to handle) is commercially liquified to a pourable lyotropic phase by adding about 25% of an alcohol–water solution. Aerosol® OT is a typical amphiphile; that is, the COONa group is strongly polar and *hydrophilic* (lypophobic), while the double peglike chain is strongly lipophilic (hydrophobic), as shown in Figure 7.21(2).[24]

The basic unit of amphiphile is the *micelle,* a cluster of polar molecules with single or double chains with their polar groups oriented in the water. Figure 7.22a shows a micelle in the water of a dilute, isotropic, colloidal solution. Figures 7.22b, c, and d illustrate the chief kinds of birefringent lyotropic mesophases. In Figure 7.22b the term *neat* comes from soap technology referring to smectic molecules structured in a lamellar phase, as a double layer, tail-to-tail with the polar groups facing the layers of water. The resulting birefringence is positive, but the strength may be weakened by some negative form of birefringence.[24]

In the soap industry, middle phase (M_1) pertains to single-chain molecules grouped into rodlike micelles of indefinite length arranged side by side in a hexagonal pattern, as shown in Figure 7.22c. In each rod the average orientation is radial with the polar groups outside to-

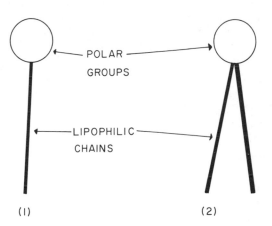

FIGURE 7.21.   Types of amphiphilic molecules (schematic).[24] (After Hartshorne and Stuart, *Crystals and the Polarising Microscope,* Arnold, 1970.) Courtesy of Microscope Publications, Ltd.

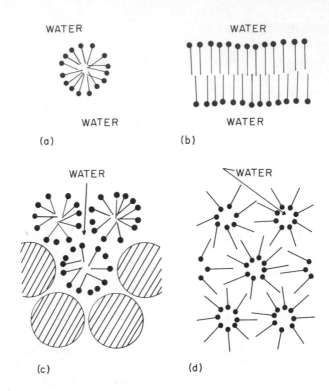

FIGURE 7.22. Main types of micelles in amphiphile–water systems. Only molecules with single lipophilic chains are shown but peg-shaped molecules would be arranged similarly, though such molecules are not known to form M_1 phases: (a) diametral section of spherical micelle, (b) smectic layer in neat phase, G, (c) middle phase, M_1, showing cross section of micellar rods (according to Luzatti), (d) inverse middle phase, M_2, showing cross section of micellar rods (according to Luzatti).[24] (After Hartshorne and Stuart, *Crystals and the Polarising Microscope*, Arnold, 1970.) Courtesy of Microscope Publications, Ltd.

ward the water between the rods. The optical sign is *negative* (in accordance with a planar structure), but the crystal birefringence is probably lessened by some degree of form-birefringence.[24]

The *inverse* (M_2) phase is in response to a higher concentration of the long molecules which then turn around so that their polar groups face water now contained in hollow rods, as shown in Figure 7.22d. This phase is also optically *negative*.[24]

Lyotropic phases can display spherulites and fan shapes. They usually are *positive*; that is, the slow component is radial. However, in neat phases the isogyres are merged as in Figure 7.23a, four of them producing a larger, *negative* (radially fast) spherulite. A pinwheel effect (Figure

7.23b) is apparent when the stage is rotated. These effects were not noticed by Rosevear (1954) in spherulites of the middle phases.[25] He had noted that the neat phase had a greater birefringence than the middle phases. In applying Rosevear's observations as a test of amphilic types, retardation must be reduced to birefringence by taking thickness and orientation into consideration. The Rosevear test does not apply to peg-shaped (double-chain) molecules. Incidentally, the sign of a spherulite is not necessarily the same as the intrinsic optical sign of the mesomorphic structure; the sign of a spherulite is merely an indication of the arrangement of the units in the sphere or disk.[24]

There are *apparently* isotropic phases in some lyotropic systems. For example, Winsor and Rogers[26] evaporated a drop of 30% solution of Aerosol® OT in water and observed an isotropic phase between an M_2 (−) and a G (+) phase. Whereas such a phase may not have an ordered structure, it may have exactly compensating positive and negative structures, possibly including some form-birefringence.[24,25]

When emulsions of water, soap, and a hydrophobic phase are frozen, they sometimes break and sometimes do not break.[27] It depends on whether or not the wall of third phase[1] around each droplet is broken by the growing crystals of either major phase. On the basis of modern knowledge of mesophases of soap and other emulsifying

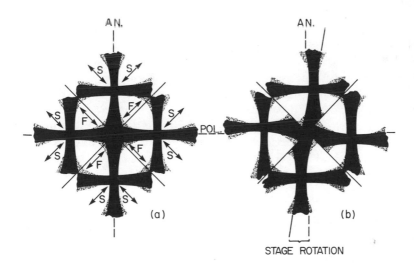

FIGURE 7.23. Merging of positive spherulites in neat phases of soaps and detergents. (a) The four positive spherulites combine to give an apparently negative spherulitic region in the center (radial direction "fast"). (b) On rotation of the stage from the position in (a) a "pinwheel" effect is obtained at the center.[24] Courtesy of Microscope Publications, Ltd.

agents, the breaking of such emulsions upon freezing should be studied on the basis of mesomorphic structure and habit at the prevailing temperatures and gradients.

7.13. SUMMARY

Microscopists often make use of crystallography because crystals are so distinctive and have so many precise characteristics that they are of enormous help in identifying and studying substances. Crystallography is an extensive subject with a formidable vocabulary, but students who have never encountered it before can take comfort from the fact that only six crystal systems are needed to embrace the (literally) millions of known pure solid phases, and that the descriptions of them are logical and orderly. An ability to visualize solid geometric shapes helps.

Crystal axes are imaginary lines employed to describe the structure and symmetry of crystals. The simplest crystal system, the isometric, is represented by three mutually perpendicular axes, exactly like the familiar x-y-z Cartesian coordinates. The structural units of such a crystal, be they ions or atoms or molecules, are arranged in equidistant fashion along these three equivalent axes. Hence the optical (and other physical) properties are the same in any direction within the crystal, which is isotropic as well as isometric. Such a crystal will appear dark in all orientations when viewed between the crossed polars of a polarizing microscope. In Winchell's very useful lists of 935 commonly encountered crystalline solids, only 1.6% of the organic and 17% of the inorganic substances are isometric, representing 94 of the 935 solids. Hence the fact that a crystalline solid is found to be isometric immediately limits the possibilities. If the index of refraction is then determined, this further limits the choices within Winchell's list. Some judicious chemical tests under the microscope will then lead to a tentative identification of the phase, and the choice can then be verified by preparing a few characteristic derivatives under the microscope and checking their optical properties.

The second crystal system is the tetragonal, which also has three axes at right angles, but one of them has different spacings along itself and is considered to be of different unit length. Viewed along the unique axis (the c-axis) a tetragonal crystal appears to be isometric because its other two axes (the a-axes) are equivalent, but viewed along an a-axis such a crystal will be birefringent because there will be two different velocities of light along the two nonequivalent axes. The third system, the hexagonal, has three equivalent axes at 60° to each other, all perpendicular to the longer (or shorter) c-axis. Hence a hexagonal crystal is also isometric when viewed along the c-axis, and is birefringent in any other position. Tetragonal and hexagonal crystals together are called *uniaxial* because they have only one unique axis. They account for 19.9% of the inorganic substances in Winchell's lists, and 11.5% of the organic substances. In appropriate orientations and with conoscopic polarized light, uniaxial crystals produce a characteristic interference figure (a Maltese cross centered on a series of colored diffraction rings) which can be observed at the back of the objective.

The remaining crystal systems have more than one unique axis. The orthorhombic system has three axes (a, b, c) all at right angles but all of different length, so that its crystals are always birefringent, in all orientations. The monoclinic system has two axes (a, b) at right angles, but the third (C) is at some other angle to the plane of the first two. The last system, the triclinic, has three unequal axes all at angles other than 90° to each other. These three systems together make up 62.1% of the inorganic and 83.9% of the organic substances in Winchell's lists. Their identification is more difficult than that of isotropic and uniaxial crystals, but determination of the angle between any two axes helps, and the characteristic interference figures also help. The refractive indices are determinative; likewise for the *sign* of the birefringence. Even without a quantitative determination of refractive indices, the sign can be determined by using a retardation plate or compensator.

Besides their six kinds of geometric structure, crystals have characteristic morphologies by which they may be recognized under the microscope. Within each system the crystals may be in the shape of flat plates (lamellar), or thicker tablets (tabular), or more regular solids with edges more or less equal (equant), or elongated prisms (columnar), or tiny needles (acicular). No matter what the morphological type, a given pure phase will always have the same refractive index (or indices) and the same optical sign, because its internal structure remains constant. Besides the five general morphological types within each system, there are characteristic shapes and faces (forms) by which the microscopist recognizes crystals. Thus isometric crystals often show up as cubes, octahedra, or tetrahedra, if they have had an opportunity to grow regular faces and sides. (If crowded together, as when crystallized from a melt, or if crushed or powdered, the pieces of any crystalline type are *grains* rather than regularly shaped units, but have the same optical properties as the beautiful equant shapes.) Well-developed tetragonal crystals are rectangular prisms and tablets, sometimes with sloping end-faces. Hexagonal crystals may be the familiar hexagonal prisms, as in quartz, or they may be the equally familiar rhombohedra (hemihedral hexagonal bipyramids), as in calcite. Orthorhombic crystals are often tabular, as in the familiar form of sulfur. Triclinic crystals have the "leaning prism" look, as in the familiar bright blue pentahydrate of copper sulfate.

The faces (forms) of any crystal habit are best defined in terms of the *Miller indices*, which are the reciprocals of the intercepts of a particular face with the crystallographic axes. Thus the faces of a cube form are represented by the indices 100, 010, and 001, while any face of an octahedron is designated 111. The Miller indices of the equant habits of all six crystal systems are indicated in Figure 7.1, and the indices of lamellar, tabular, columnar, and acicular crystals of the same system are the same as those shown there.

Most of the usefulness of crystallography to the microscopist comes from long experience in recognizing particular systems from the way their crystals of characteristic habit appear as they lie on this face or that, or as they develop from solution on a slide, or as they freeze from a sample held on a hot stage or cold stage. The text contains dozens of tips on how to do this, and is illustrated with many examples, but the microscopist must familiarize himself with all of the techniques and endure all the pitfalls before he can gain the facility that comes with that experience. The systematic study of pure compounds, and especially the faithful recording of all optical properties of every substance that comes to his attention (under some system that allows rapid retrieval of the information), will do wonders toward reaching that state.

Finally, we discussed liquid crystals, which are manifestations of residual order held over as a solid begins to melt or to dissolve in a solvent. The order usually arises from the persistence of oriented electric dipoles, which is why liquid crystals can be manipulated by electric fields to display digital readouts in calculators and watches. The narrowly limited temperature range of thermotropic liquid crystals is a reason why they are used also in digital thermometers.

Photomicrography

8.1. PHOTOMICROGRAPH: IMAGE PRODUCED BY LIGHT, ELECTRONS, OR X-RAYS

A *photomicrograph* is a developed image produced on a sensitized surface by means of radiation emerging from a microscope. The radiation may be visible light, or ultraviolet light, or a beam of electrons, or x-rays. By common agreement, a photomicrograph is an enlarged image, with a magnification of 10× or more.[1] If less than 10× (as could be accomplished by a camera with special lenses, but no compound microscope), the picture is properly called a *photomacrograph*. Just to keep the terms straight, a *microphotograph* is a microscopically small photographic image,[2] such as a microfiche of a page or more.

8.2. TWO ATTITUDES: ARTISTIC VERSUS SCIENTIFIC

There are at least two *attitudes* toward photomicrography and, indeed, all of photography: the artistic view and the scientific view.[3–5] The artistic view may be that of the *fine arts*, whereby a beautiful photograph is created in the eyes of the photographer and, it is hoped, in the eyes of the observer. Beautiful color effects, for example, may be obtained by placing appropriate thicknesses of cellophane tape in a diagonal azimuth between crossed Polaroid® polarizing sheets, or by photographing any appropriate anisotropic specimen between crossed polars. A pretty photomicrograph obtained in that way might also have value in the *graphic arts*, as in advertising photographic equipment or polarizing film or cellophane tape, or even in advertising soap or any other material that is crystallized from fusion. Unless the purpose is to advertise polarizing film, it may be just as well to obtain beautiful color effects simply by illuminating some crystals[6] at various angles by differently colored beams of ordinary light. In other words, the photographic approach in art is subjective.

However, in science and technology the approach must be objective. Photographs vividly describe objects and observations (Figure 8.1) as recorded in records, reports, publications, specifications, patents, and other communications, and are relied upon implicitly by the observer. Generally, graphic descriptions of these types are interpretive illustrations of what has been observed repeatedly, or characteristically. Such photomicrographs can illustrate either an average or an extreme. Like caricatures they can exaggerate, for better or for worse. They are pictures taken of "posed" specimens, and compete with drawings, diagrams, patterns, and spectra.

Another purpose of photography is to record a series of changes in a

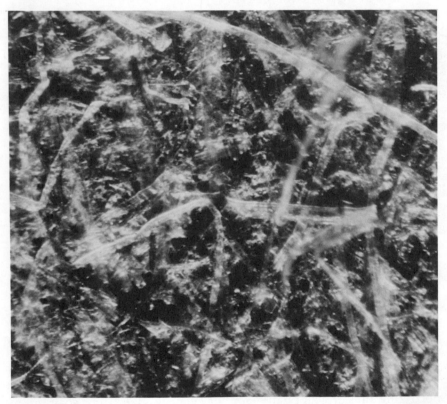

100 μm

FIGURE 8.1. Surface of wallpaper stained so that chemical pulp appears white and ground wood appears dark. The photomicrograph tells a story.

specimen, such as those which occur during cooling, heating, extruding, stretching, relaxing, or immersing.[7] The intervals between exposures may be such that a series of snapshots may be taken, preferably on roll film, or possibly with a stack of holders loaded with cut film.[8] For faster sequences, ordinary or high-speed cinematography[9] is required to record the motion. For slow sequences there are devices for taking time-lapse photographs at appropriate intervals.[10]

In electron microscopy, especially that by transmission (rather than reflection) of electrons, the specimen is heated by the electron beam and may change in appearance. To show this, or to get the most in fidelity, the usual attitude is to take as many photomicrographs as possible in a short time, and to study them rather than to study the specimen visually (Chapters 12, 13, and 14).

8.3. EXPERIENCE: RECORDS OF NEGATIVES

Experience is the only way through which to achieve satisfactory results in photomicrography. Keep a good record of each photographic exposure on some sort of data sheet (or on an envelope which is to contain the negatives), such as is illustrated in Table 8.1. It is especially important to record the reading of the exposure meter along with the time of exposure. By correlating these with negative and positive results (to be written in "Remarks" column, Table 8.1), one gains enough experience to avoid having to take more than one frame per specimen to ensure success.

8.4. IMAGINATION

Imagination is the formation of a mental image of something not yet perceived. In microscopy, imagination helps in selecting the best combination of all the other attributes contributing to visibility. In photomicrography, imagination serves in selecting the best conditions, equipment, techniques, and photosensitive materials to record the image both artistically and scientifically. Above all, imagination helps to combine the best of the old and the most promising of the new.

8.5. RESOLVING POWER

Resolving power in photomicrography, as in visual work, is expressed by the numerical aperture (NA). In photomacrography and pho-

TABLE 8.1
Example of a Record of Photomicrography on Roll Film

Date _____

Experiment or Project _____

Film type _____   ASA _____

Frame	Specimen	Location	Mounting medium	Objective	Eyepiece	Magnification	Illumination	Exposure meter	Filter	Exposure		Remarks
1												
2												
3												
4												
5												
6												
7												
8												
9												
10												
11												
12												
13												
14												
15												
16												
17												
18												
19												
20												
etc.												

FIGURE 8.2. Vertical camera with specimen stage. The vertical camera is being used with a short-focus photographic lens. Note that the bellows length may be fixed on the flat, vertical rail, allowing the fixed bellows to be moved vertically or radially on the round upright. The specimen stage is fitted with an upright rack-and-pinion adjustment for focusing, and two horizontal ones for selecting the field to be photographed.[11]

tography, resolving power is expressed by the f number (focal length divided by the actual opening to the lens). By neglecting the difference between the tangent, i.e., the f number and the sine of half the angular aperture, the practical conversion is $f = 1/2NA$. This estimate is adequate for a compound microscope in which the tube length is fixed. But, when photographic single lens systems (simple microscopes) are used with a bellows (Figure 8.2), the f number varies with the ratio of image distance to the object distance (magnification), as shown in Figure 8.3. As the object distance (D_o) becomes smaller in focusing to a larger image (I), the angular aperture (α) becomes greater; therefore the NA becomes larger.

8.6. RESOLUTION BY PHOTOMACROGRAPHIC LENSES

Resolution by means of photo*macro*graphic lenses is as good or better than by means of compound microscopical objectives because the *macro*lenses[11,12] are so highly corrected for a wide field. They are particularly good at bellows extensions of 250 mm or more, provided that they have been mounted for close-up objects (the reverse of the situation in cameras when focused on relatively distant objects). For close-up work, the more convex of the two outer glass surfaces is mounted toward the object; for distant objects, the more convex glass surface is mounted toward the photographic plane.[11] Macroobjectives of focal lengths 32

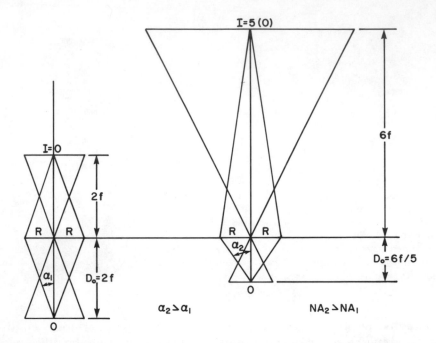

FIGURE 8.3. The NA of a macrolens system increases with magnification.[12] Courtesy of Microscope Publications, Ltd.

mm or shorter are usually mounted with the standard microscopical thread.[13] If not, an adapter is usually available for the lens board shown in Figure 8.2.

Because macrolenses are so well corrected out to their extremities and because their characteristics are not modified by an eyepiece or projection piece, they can be used to an advantage with a very long bellows, using an extension bellows or even a dark room. Indeed, pictures taken this way represent a kind of photographic art.

Compound light *microscopes* function somewhat differently in photomicrographs than in visual work.[12] Instead of forming a virtual image by means of the eyepiece and the eye (Figure 8.4a), a real image is formed in the photographic camera. The eyepiece may be left in place, but the objective is raised to form its real image *below* the focal plane of the eyepiece, which now acts as a projecting piece so as to enlarge the first real image into a second, much larger, real image either on the ground glass or on the photosensitive material, as shown in Figure 8.4b.[14] While either the Huygens or the Ramsden eyepiece may produce a satisfactory photomicrograph if the bellows (projection) distance is close to 250 mm, it or any other eyepiece is not designed for a projection distance much different from this. However, there are projection pieces built much like an Huygenian eyepiece except that the "eye" lens can be moved toward or away from the field lens by means of a helical screw with an indicator to show the proper position for a given projection distance.[11] An amplifying lens is even better for photomicrography. It is effectively a concave lens which *diverges* the rays to focus in the plane of the photograph, as shown in Figure 8.5.[14]

The main purpose of amplifiers originally was to correct for curvature of the image; their main disadvantage is that they do not produce a virtual image and so cannot be used directly with the eye. Today, special flat-field achromatic objectives are available. They contain a large number of glass components and are expensive, but they are designed for use with eyepieces in both visual and photomicrographic work.[14] Yet for attaching cameras such as those that use 35-mm film, now so popular, the flat-field objectives with their eyepieces still leave something to be desired.[15] Such shallow cameras crop the field so much that 8 × 10 in. prints cannot be made to take full advantage of the flat-field objectives and wide-field eyepieces. "Transfer" lenses help to some

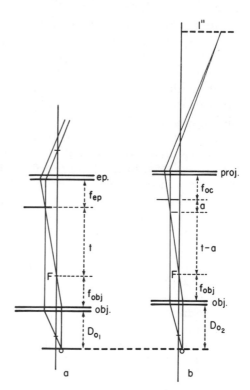

FIGURE 8.4. Virtual image for visual observation (a) and real image for photomicrography (b). In the latter case, the intermediate image is formed at the shorter distance $(t-a)$ from the objective.[14] Courtesy of Microscope Publications, Ltd.

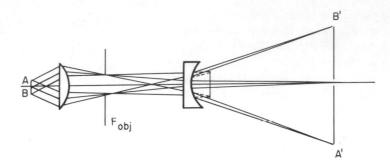

FIGURE 8.5. Formation of the real image for photomicrography by objective and amplifier.[14] Courtesy of Microscope Publications, Ltd.

extent, but not completely. As shown in Figure 8.6, Quackenbush[15] proposes the use of an extension tube with the commercial "transfer lens." Another modification (not shown) is to open up the diaphragm in the eyepiece. Quackenbush now suggests the manufacture of lower-power eyepieces for use with shallow cameras: a 2.5× eyepiece for "standard" microscopical tube (160–170 mm) and a 2.0× eyepiece with a tube-length factor of 1.25.

If a camera such as that used for cinematography has a nonremovable lens, its focusing indicator should be set at infinity.[14]

Resolution in a picture, as a response to a given resolving power, depends upon imaging two points in the object on two grains or dots of silver halide in the film that are separated by at least one intermediate grain or dot. After exposure to light these grains are developed (reduced) to silver. The resulting graininess[1] is what determines the resolution in the positive or negative.[2] In color photosensitive materials, the grains involved in resolution may be pigments or dyes, but the same principle holds. In halftone processes for press-printing of pictures, a grid of pits has been etched into the lithographed plate. These pits are to hold the ink; they (or the dots functioning in photo-offset processes) define the resolution obtained in the mass reproduction. Obviously, that resolution is far less than that of the original photograph.

8.7. CONTRAST: NATURE OF PHOTOSENSITIVE MATERIALS

Contrast in a photomicrograph, more so than that obtained in the visual image, depends on the nature of the photosensitive materials, the kind and concentration of developer, and the temperature and time of development.[16] While there are many materials to choose form (and experimentation is certainly advocated), unless experience is to the contrary, a good type of black-and-white film in roll or sheets is exemplified

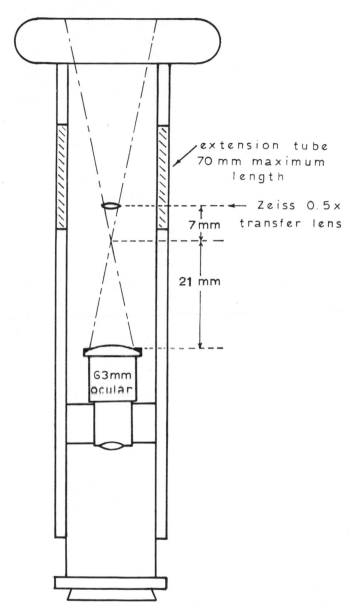

extension tube
70 mm maximum
length

Zeiss 0.5x
transfer lens

7 mm

21 mm

63mm
ocular

FIGURE 8.6. Quackenbush modification of a 35-mm, wide-field camera assembly.[15]

by Kodak Tri-X® panchromatic, with an American Standard Association (ASA) film speed of 400. Kodak developer D-76, diluted 1:1, represents a very good all-around developer. For finer-grain film Kodak Microdol-X® is typical, but it will require $\frac{1}{2}$–1 f-stop more exposure.[17] Development in a tank (loaded in *complete darkness* if panchromatic) does not require a shortstop bath, if the fixer is to be discarded after use. Agitation is maintained during fixing to obtain maximum "brilliance."[17]

Contrast may be increased by changing to slower negative materials such as are used in spectroscopy, projection, copying, and metallography.[16,18,19] However, if the speed of Tri-X Pan® is needed, as with particulate, anisotropic specimens between crossed polars, a more alkaline, more contrasty developer can be used.[10,18,19]

Contact positives, even the small ones of 35-mm film, serve two important purposes: (1) they allow each frame to be examined carefully for choosing the best negatives to be enlarged, and (2) they provide a basis for filing the negatives. For such purposes a print should be given full time of development (90 sec) and full inspection under normal white illumination.[17] Kodak Dektol® diluted 1 : 2 is excellent for developing prints. Follow directions[18] and use a Kodak OC® safelight that covers the whole working area.[17] A shortstop bath of 28% acetic acid (glacial, 99% acetic acid poured *cautiously* into cool water at the ratio 3 : 8) or Kodak® Indicator Stop Bath (yellow at the start, dark purple when spent) is used for 20 sec between developer and fixer.[17] Do not rinse with water between shortstop and fixer.

If economy is foremost, the fixer can be based on commercial bulk sodium thiosulfate ("hypo") crystalline hydrate ($Na_2S_2O_3 \cdot 5H_2O$), 2 pounds to make 1 gallon of solution. Dissolve in about 3 quarts of water by mechanical stirring. Then add 8 ounces of Kodak® acid hardener or equivalent[19] hardener to toughen the gelatin film. Add water to make 1 gallon. If convenience is foremost, packaged formulated fixer can be stored ready for instant dissolution. In either case, fix for at least 10 min, or longer if necessary to clear the film. Before each session, test the fixer with a commercial test solution.[17]

Aim for correct exposure, but err on the side of overexposure. There is not much to be done with negatives which are too thin. But if overexposure is not severe, the negative may be lightened with Farmer's reducer, a 20% solution of "hypo" ($Na_2S_2O_3 \cdot 5H_2O$) to which is added (just before use) enough potassium hexacyanoferrate, $K_3Fe(CN)_6$ to give a lemon-yellow color to the solution. Negatives should have been fixed and washed first. If dried, too, soak the negative in water before reduction.[19] Cut roll film into frames and experiment with one frame at a time in a white tray. Use white light to observe the reaction, while gently rocking the tray. Just before the optimum reduction, plunge the negative into a bath of wash water. If more reduction is desired, repeat the operation, after dissolving more potassium hexacyanaoferrate, if necessary. Reduction of a slightly overexposed negative is an excellent way of *gaining contrast.*

Very contrasty transparencies for projection can be made by first slightly underexposing Kodak Panatomic-X® film (ASA 32) and developing *directly* to a positive transparency by employing a Kodak direct-positive film-developing kit and following directions carefully. The too-dense transparency is then reduced with Farmer's solution to the desired transparency and contrast.

Contact prints of negatives are made with a light-box printer. Enlarged prints are made individually with an enlarger. The two methods require different types of photosensitized paper. There are many types, but with all, the contrast is designated by numbers from 1 (or 0) through 5 (highest contrast). Aim for a negative of average contrast and density, that is, one that will print satisfactorily on a No. 2 paper. An average negative allows for extra contrast by employing a No. 4 or 5 paper.

Resin-coated positive photosensitive material, such as Kodak Polycontrast Rapid RC® printing paper, can be fully washed in 4 or 5 minutes and dried almost as quickly.[17]

Kodak Panatomic-X film (ASA 32) is also a good negative black-and-white film to use in trials for determining conditions to be used with color film, such as Kodachrome (ASA 25). Kodak high-speed Ektachrome® (ASA 125) is especially good for making 2 × 2 in. slides of polarization colors. By obtaining special processing, this film can be exposed at even higher speeds[20] and still produce good transparencies.

8.8. CORRECTIONS FOR ABERRATIONS

Corrections for aberrations in photomicrographic equipment have been made almost perfectly for projection distances of 250–500 mm. As discussed in Section 8.6, a serious problem of aberrations comes with attachment cameras, including those for 35-mm roll film, which have a projection distance of less than 250 mm, and often only one-third of this. All such cameras should be equipped with an auxiliary "transfer" lens (of about 8.3 mm) so as to form a sharp real image in the plane of the film. Motion-picture cameras used for cinemicrography should be equipped with an auxiliary lens of commensurate (shorter) focal length.[14]

8.9. CLEANLINESS IN THE DARKROOM

Cleanliness, care, and accuracy in photomicrography are very important. The development and fixing of photographic images are chemical processes, with reaction rates dependent on the temperature. If the temperature is one degree centigrade too high, there will be a 10%

overdevelopment and the negative may be too dense to print even on the least-contrasty grade of printing paper. The time of the developer is equally important. Agitation of the solution should be in accordance with instructions. In all operations, these three factors (temperature, time, and agitation) should be kept constant so that results are reproducible.[17]

All film should be handled carefully. Wet film is plastic and it replicates fingerprints permanently. The films should be dried dust-free. Printing should be done only with completely dried negatives because only these are clear and transparent through and through. Positives and negatives should be carefully and neatly marked. Only soft pencils should be used to mark the backs of prints, lest the pencil marks may show through.

Darkroom processes necessarily involve many liquids. All spills should be cleaned up as soon as possible to protect your own work. The dust from dried fixing solution will ruin all new film by producing tiny white spots on it. If you share the darkroom with others, your responsibility to them is obvious. By the same token the absent darkroom workers owe it to one another to remove all finished positives and negatives and to leave the work space clear.

Good housekeeping and fair play are small prices to pay for the advantages of darkroom facilities as compared with sending photographic work out or using instant, automatically processed materials such as Polaroid.® Not that such materials are not valuable for their inherent advantages. These are tremendous boons to the busy microscopist who can satisfy a customer by snapping a picture and handing it out within a few minutes. However, sooner or later the customer or his relatives or the microscopist himself will want copies, and without loss of resolution or contrast. Even by use of instant, automatically processed negatives, the inherent lack of control and the variation of sizes, shapes, or finishes of opacities or transparencies may transcend the convenience of speed, and therefore call for an orderly, clean, versatile and ready darkroom.

The photomicrographed field of a specimen must be clean of dirt, fingerprints, artifacts, and confusing associations. Otherwise, the extraneous parts or particles are very likely to be misinterpreted. If any such artifacts crop up in the picture, they should promptly be cropped out.

8.10. DEPTH OF FOCUS

Depth of focus is much more critical on photosensitive materials than on the human retina because more attention must be paid to the optical system and its adjustment. Numerical aperture may have to give way to depth of focus by a reduction of the condenser's aperture. If this maneuver alone is unsatisfactory, try an objective of lower NA or, better, use an iris diaphragm that comes with or can be inserted into the objective of high NA. If satisfaction is not yet achieved, try making the specimen thinner without sacrificing the pertinent third dimension. As a final resort, multiple micrographs may be required both to map the gross structure and to resolve the fine detail.

8.11. FOCUSING

The best *focusing* of an image to be photographed is done on a translucent or ground-glass plate or sheet in the exact plane where the photosensitive material will be exposed. The ground surface of such a ground-glass plate should face toward the microscope in place of the emulsion. After visual examination of the marked area representing the size and shape of the photosensitive area, the center of the area at least should be focused with the aid of a magnifier commercially available for the purpose. A clear glass pane substituting for the ground glass will enable the photomicrographer to move the magnifier all around the area. Usually this is an area of 35 in.^{2}, because 5×7 in. is a good size negative for publication, for exhibition, and for direct visual examination. Inspecting all that area each time would be tiresome, but fortunately once the proper optics are obtained to get a flat field, and all specimens thereafter are prepared flat and with their surfaces perpendicular to the microscopical axis, a cover glass (large enough for the magnifier) may be cemented over the center of the rough side of the ground-glass plate for checking the focus, so that the plate does not need to be interchanged with a clear pane each time.

Cameras taking roll film (such as 35-mm, Polaroid®, or motion-picture film) usually do not have provision for direct focusing on a translucent screen. Such cameras generally have a withdrawable observation tube fitted with a reflecting prism on one end and an eyepiece on the other. The eyepiece is focusable on a target, or there is some other means of assuring that when the object is in visual (virtual) focus the real image to be photomicrographed will also be in focus. The positioning of this target or other indicator is the responsibility of the manufacturer, but the visual observation is the responsibility of the photomicrographer. His eye, like that of the microscopist doing direct visual work, should be focused at infinity, that is, his eye muscles should be relaxed. If the image is out of focus, he should use the focusing mechanism and *not* his eye's accommodation. For checking on new equipment or an out-of-focus micrograph, a magnifier with a small ground-glass plate mounted in the proper plane can generally be put in place of the film while making adjustments.

Accurate focusing is just as important with an enlarger as it is with a microscope. The image should be in sharp focus on the easel in the exact position for the actual photocopying. A satisfactory magnifier, usually with a 45° prism for viewing at an angle to the easel, should be used rather than checking with the naked eye. Remember, too, that the focusing mechanism of an enlarger also changes the bellows length and *may* change the magnification significantly.

Exact focus is not always the best adjustment for a microscope, because boundaries and other lines may be too fine to see. Slight over- or under-focusing often broadens lines and increases contrast. This is true when using either transmitted or reflected light, or even an electron microscope.

8.12. ILLUMINATION

Illumination, above all, should be uniform in photomicrography. With a large-frame camera, such as 5 × 7 in., a probe connected to a photometer[21] may be used to examine the area for variation in intensity. Lack of uniformity of illumination usually comes from an off-center illuminating beam. With Köhler's illumination, centering the beam is easy because the lamp's (field) diaphragm is imaged in the plane of the specimen in focus. The field diaphragm is contracted and the substage condenser is focused to bring the field diaphragm within the field of the specimen by means of the microscope's mirror. If the beam is still off-center, move the condenser about by means of its centering screws, if any are present. Even when some of the condenser's aperture is screened out, the other two azimuths can be evened out with Köhler's illumination. Ground glass introduced anywhere in the illuminating system interferes with Köhler's illumination.

Unevenly illuminated negatives to be enlarged can be evened to some extent by dodging: quickly moving a hand back and forth in the brighter portion of the image during exposure of the positive print. In contact printing, one or more layers of translucent paper torn to shape and placed over the negative can help salvage an uneven negative that cannot be replaced.

8.13. SPECTRUM OF EFFECTIVE RADIATION

The spectrum of *radiation* (wavelengths of light) that affects photosensitive material is much broader than the visible spectrum. Heat radiation (particularly from that of an arc lamp) can be actinic to panchro-matic photographic film, and can be harmful to specimen and microscope. A cell of water is of some help to filter out infrared radiation, with a piece of heat-absorbing glass immersed in it.[19] Ultraviolet light is very actinic. It should be filtered out to avoid overexposure and chromatic aberration. Ordinary glass will take out the shorter wavelengths, but a UV filter is better.[19]

Nowadays photomicrography in color is commonly desirable, es-pecially to record polarization colors. With the intensity of incandescent tungsten lamps controlled by varying the voltage, which varies the color, a daylight filter should be kept in use so that exposure meters may be read more accurately.

Mercury arc lamps and other high-intensity sources of specialized radiation should be supplemented with a proper filter for the sensitive material, especially when using orthochromatic or "commercial" film. If UV light is used for fluorescence or absorption by the specimen, be sure to protect the eyes with UV-absorbent goggles and to use UV-transmitting optics where required. If the UV radiation is to be photo-micrographed, the microscopical objective should be corrected also for a wavelength in visible light (e.g., green, 440 μm),[19] so that the specimen can be focused visually and photomicrographed in the UV.

8.14.   ANISOTROPY

Anisotropy in a specimen is commonly portrayed in color by observ-ing the specimen *between crossed polars*. The term "under polarized light" is ambiguous because it does not indicate whether or not there are two polars, and whether or not they are crossed entirely or partially.

The main problem with making randomly oriented, anisotropic particles visible between crossed polars is that some such particles are in positions of extinction, and so not all are visible. The problem cannot be solved by rotating the stage because some particles at least will be moving off-center. The solution involves synchronously rotating the crossed polars with respect to the stationary particles. At low powers the device shown in Figure 5.13 is good for simultaneously rotating the crossed polars by hand. For higher powers with a petrographic micro-scope, a bar connecting the rotatable polarizer and rotatable analyzer will keep them in synchronism.[11]

8.15.   USEFUL MAGNIFICATION IN PHOTOMACROGRAPHS

Useful magnification in photo*macro*graphs taken with a *macro*lens (photographic simple microscope) is more than 1000 × NA, which is the

FIGURE 8.7. Photomacrograph, taken simultaneously, of a piece of plastic and a centimeter scale recording the magnification.

approximate limit of useful magnification (LUM) with a compound microscope. The reason, as explained in Section 8.5 and in Figure 8.3, is that the angular aperture is increased significantly as the object distance is shortened to focus the enlarged image at the increased magnification.[12]

In Table 2.2 the limiting distance between points is 0.15 mm for the unaided eye 250 mm away from the object. However, for a picture hung on a wall to be viewed at about a meter away, the minimum distance between points should be 4 × 0.15 = 0.6 mm. Yet if the micrograph is to be projected to, and viewed from, a screen, the minimum distance between points *on the micrograph* should be

$$\frac{\text{viewing distance}}{\text{projection distance}} \times 0.15 \text{ mm}$$

Actual magnification of a micrograph should be indicated by a scale that will be reduced or enlarged along with the micrograph. In low-power work such as is illustrated by Figure 8.7, magnification can be computed from a scale in the field of view, as photographed simultaneously with the object. In photomicrography with a camera that can be provided with a ground-glass plate exactly in place of the photographic film, an image of a standard stage micrometer can be measured as enlarged and focused on the ground glass. In contact printing a commensurately enlarged transparent scale can be printed alongside the negative. Then if the negative is enlarged or reduced in printing, the scale will be changed exactly in accordance. Alternatively, two distinct points on the negative can be measured linearly there and also on the positive, and the ratio can then be applied to the magnification measured on the ground glass before exposing the negative. With roll film, one frame should be used to photomicrograph a stage micrometer *for each set* of magnifying conditions. The relevant negative scale is enlarged commensurately with each different enlargement of the one micrograph, and the average distance between smallest divisions on the negative image of the stage micrometer will give the information to make a scale on the particular positive micrograph. Be sure that each different enlargement is marked at least with the accurate magnification on the back, if not by an accurate unit scale on the front.

8.16. FIELD OF VIEW

The *field of view* on a finished micrograph should show as much of the field as is required to fulfill the purpose, no more and no less. If you are cropping a micrograph by drawing rectangular lines or a circle on it, be sure the lines are permanent. Grease pencil is fine for temporary markings, but they are easily rubbed off with a finger. Use ink, lest someone else do the cropping incorrectly.

Each micrograph should have a single, separate, vivid purpose. If you intend to cut a field into parts, do it before you release the micrograph to others, especially to publishers and photoengravers. If your micrograph has a top and bottom, a left and a right (and most do), indicate those definitely and permanently on the print.

8.17. SUMMARY

Photomicrography is the art of photographing images produced by a microscope in order to provide a record for display, for measurement, for publication, or for future reference. It combines all the aspects of photography with all the principles of microscopy discussed in the previous chapters, both in the matter of equipment and in the techniques to be employed. The aim is a clear, bright picture of the details chosen by the microscopist, large enough to serve its intended purpose but with sufficient resolution to show the desired detail, and in focus out to the edges. If in black and white, proper contrast is necessary to reveal that detail; if in color, there must be due regard for correct color balance and color density. Since a camera does not have the depth of focus of the human eye and cannot accommodate itself, extreme measures must be taken to obtain a flat field in the plane of the film. All this makes photomicrography more exacting than visual microscopy, but the results are worth it in that a good picture can convey more information than thousands of words, and can provide a compact way of storing results. Photomicrographs can also be beautiful in their own right, quite apart from their scientific content, and many have won prizes in art shows.

In visual microscopy the compound light microscope provides the eye with a virtual image seen through the usual eyepiece, but micrography requires a real image to be formed within the camera at the plane of the photographic film. Such an image is obtained either by using a special projection "eyepiece" in place of the usual ocular, or by using a concave lens (an "amplifier") to diverge the rays so that they form a real image of the proper size at the plane of the film. There are also expensive flat-field achromatic objectives which can be used both in visual work and with a 35-mm camera. The combination is not always satisfactory because a compact 35-mm camera back has too short a "throw" to the film and too small a field to take advantage of the flat-field objective and wide-field eyepiece. Although contact prints made for 35-mm film are easy to store and provide a quick survey of the exposures taken, they must always be enlarged singly and laboriously to obtain prints large enough for direct inspection or for publication. Furthermore, most 35-mm cameras are "blind" and the image cannot be focused visually over the entire field at the plane of the film. Perhaps only in making 2 × 2 in. color slides for projection does the 35-mm camera come into its own, and even there the limitations of field and focus still hold.

A good objective of proper NA to achieve the desired resolution, used with a transfer or amplifier lens and adjustable bellows, will provide a field large enough to examine the image in detail and to take negatives which can be contact-printed and still be large enough for reports and publication. The 5 × 7 in. format is a convenient size and is economical of photographic materials.

The beginner will find it helpful to start with Kodak Tri-X® panchromatic film developed in Kodak® D-76 and fixed in a 25% solution of

$Na_2S_2O_3 \cdot 5H_2O$ containing an acid hardener. All exposure times, light intensities, and development times and temperatures should be recorded along with microscope settings and selections, in order to build up experience. Nothing can be done about underexposed negatives, but overexposed ones can be lightened by soaking in a solution of $K_3Fe(CN)_6$ and $Na_2S_2O_3$. Fortunately, considerable flexibility is provided by photographic printing papers which are available in five degrees of contrast, so a flat negative can still produce a lively print if coupled with a contrasty paper. Most pictures can be improved by cropping the negative or reframing the subject. Many helpful hints for achieving excellent finished pictures are included in the text. Only by working in his own darkroom can the microscopist enjoy full flexibility and control of his photographic operations; cleanliness and order are essential there.

In the operation of photomicrographic equipment there is also no substitute for experience, but many gadgets are helpful. A ground-glass plate with a small cover glass cemented to the center of the ground surface will speed up the focusing operation, once everything is selected and aligned for a verified flat field at the focal plane. A small magnifier helps greatly in examining the image during focusing. Similarly, a 45° magnifier allows sharp focusing of the image during enlarging. There are photometers to check the uniformity of illumination, and exposure meters to measure intensity. Since there are special requirements of resolution and magnification in photomicrography, and so many personal preferences in photography, a large selection of special objectives, eyepieces, illuminators, and camera mounts has been developed. All have their advantages and disadvantages, summarized herein. The user will have to familiarize himself with the optical principles involved, as outlined in this chapter, before he can make intelligent selection and use of these accessories. Extension tubes, bellows, cameras, and camera mounts all alter the optical path length, and so will change the magnification. Hence an independent means of measuring the actual magnification is needed, and it is preferable to indicate the actual extent of magnification directly on the final print. Maximum useful magnification is governed by the limit of resolution, which in turn is set by the kind of photographic materials and by the method of reproduction used in publication.

Contrast: Phase, Amplitude, and Color

9.1. CONTRAST: COLORLESS AND COLOR

Contrast, as was pointed out in Chapter 2, is an attribute contributing to visibility which is next to resolution in importance. Indeed, two parts of an object that are resolved separately still are not seen separately unless their images are contrasted against what is between them. In light microscopy, we are concerned about two principal kinds of contrast: colorless and color. In both kinds we are also concerned with intensity, the *amplitude* of the light waves. The intensity of the colorless kind of contrast is in terms of black, white, and intermediate grays. This kind of contrast comes from interference and reinforcement of light waves which originate at each point in the object but travel different paths and lengths through the optical system to form the final image.

9.2. INTERFERENCE: DESTRUCTIVE AND CONSTRUCTIVE

Destructive interference occurs when two waves are *out of phase*.[1] In Figure 9.1, waves b and c have the same length, λ, but are exactly one-half wavelength out of phase. Since they happen to have the same amplitude, the net result is complete interference (zero intensity, i.e., darkness).[2] *Constructive interference* occurs when two waves are *in phase*. In Figure 9.2, waves b and c still have the same length, λ, but are exactly in phase. Since they also happen to have the same amplitude, the net result is constructive interference to the extent of twice the original amplitude (intensity).[2]

A diffraction image is produced by interference and reinforcement of various light waves from all points in the object (of specified

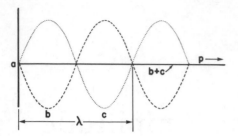

FIGURE 9.1. Two light waves, b and c, of the same length, λ, being propagated in the same direction, p, and having started exactly *one-half* wavelength out of phase. At any point along the path, p, the net result is complete cancellation (zero amplitude, 0a).[2]

thickness) along different paths within the angular aperture of the objective. Thus, by controlling the angular apertures of the condenser and objective, the microscopist can go a long way in controlling the kind and extent of contrast he obtains in the image of a given object. In bright-field illumination[1] (see Figure 2.11) by transmitted light, the control of the angular aperture of the condenser is by means of a diaphragm such as the well-known iris diaphragm of Figure 2.14. It provides a variable annular stop, as shown in Figure 9.3a.[3] With the iris diaphragm, nonscattering and poorly scattering parts or particles tend to appear brighter than highly scattering ones. The reverse is true with a central stop, which cuts out some but not all of the direct rays from the condenser into the objective, as indicated in Figure 9.3b. With this differential stop the more highly scattering parts tend to appear brighter than lesser-scattering ones. This result is accomplished most simply by inserting into the filter rack of the condenser a diaphragm with an opaque central disk (Figure 9.4) just large enough to screen out all direct rays to

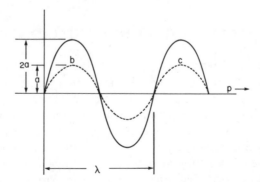

FIGURE 9.2. Two light waves, b and c, of the same length, λ, being propagated in the same direction, but having started exactly in phase. At any point along the path, p, the net result is constructive interference (double amplitude, 2a).[2]

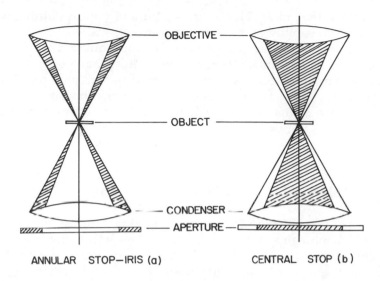

FIGURE 9.3. Reciprocal cones of bright-field illumination, illustrating use of *annular* (iris) stop (left) and complementary *central* stop (right). The shaded areas indicate portions that are cut off by the particular variable stops.[3]

the objective when the condenser iris diaphragm is wide open.[4] With this arrangement the field appears black, but those components of the object which scatter light appear as bright spots against the dark background. As a consequence, dark-field illumination is extremely contrasty, and is often very helpful in improving visibility.

If, however, the central disk of the diaphragm shown in Figure 9.4 is transparent and colored, say red, and the annulus is blue, the background will appear red instead of black, and the parts or particles that scatter light will appear blue instead of white. Such an arrangement is

FIGURE 9.4. Fixed central stop, which fits under the substage condenser.[4] The diameter of the permanent stop is selected according to the numerical aperture of the objective and whether bright-field or dark-field effects are desired.

called a Rheinberg filter.[5] It produces a kind of optical staining which improves the visibility of many transparent objects.

These are some of the many ways to obtain contrast. They should all be tried fully on any unfamiliar specimen, rather than going directly to the phase-amplitude-contrast method which has its own problems of interpretation. There are, however, times and circumstances in which contrast by the earlier methods is inadequate, and staining or treating the specimen is impossible or undesirable.

9.3. PHASE-AMPLITUDE CONTRAST

The phase-amplitude method of obtaining contrast elaborates on the use of diaphragms for the separation and recombination of direct vs. diffracted rays.[5-7] Figure 9.5[2] illustrates a typical phase-amplitude system for *transmitted* light. Köhler illumination is incident on the lower focal plane of the condenser, where there is an annular diaphragm with an opaque central stop. Rays through this diaphragm are focused as a hollow cone onto the specimen.

In the back focal plane of the objective there is a conjugate annular diaphragm which is called a *diffraction plate* because the diffracted and undiffracted rays strike different parts of it[2] (Figure 9.5). It is also called a *phase plate* because the phase relationship is altered here. If, for example, the undeviated rays are retarded by a transparent film of proper thickness, *bright* contrast results[6] (Figures 9.6a and 9.7a[7]). If the phase-delay film is on the center instead of on the annulus, dark contrast is the result (Figures 9.6b, 9.7b[7], and 9.8[7]). The actual degree of darkening or lightening depends upon the retardation of the light waves from the object. In Figure 9.6c, the retardation is assumed to be $\frac{1}{4}\lambda$ and the refractive index of the object is assumed to be slightly greater than that of the medium.[6] If, instead, the refractive index of the object is slightly lower than that of the medium, the results are reversed. With either a bright- or a dark-contrast phase plate, the annulus is usually coated with a partially absorbing (very thin) film, such as silver (Zernike) or carbon soot (Wilska).[5,7] The purpose is to reduce the amplitude (intensity) of the undiffracted direct rays, so as to be commensurate with the low intensity of the diffracted rays.

From the foregoing it can be understood that a great variety of accessories and their combinations is possible, and many are commercially available. The trend of the manufacturers has been to develop a wide variety of combinations with annuli interchangeable on the one condenser and with phase- and amplitude-modification plates built into the many objectives of various contrasts and numerical apertures. The

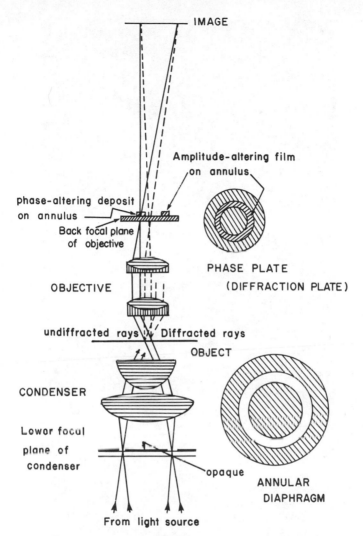

IMAGE

Amplitude-altering film
on annulus

phase-altering deposit
on annulus
Back focal plane
of objective

OBJECTIVE

PHASE PLATE
(DIFFRACTION PLATE)

undiffracted rays   Diffracted rays

OBJECT

CONDENSER

Lower focal
plane of
condenser

opaque

ANNULAR
DIAPHRAGM

From light source

FIGURE 9.5. Typical phase-amplitude microscopical system by transmitted light. In this case, the phase-altering material has been deposited in the annulus.[5]

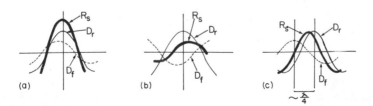

(a)   (b)   (c) $\sim \frac{\lambda}{4}$

FIGURE 9.6. Principle of phase-contrast illumination.[6] (a) Bright phase contrast. (b) Dark phase contrast. D_r = direct (zero-order) wave, D_f = diffracted wave, R_s = resultant wave. (c) Phase object (with refractive index slightly greater than that of surroundings). Courtesy of Microscope Publications, Ltd.

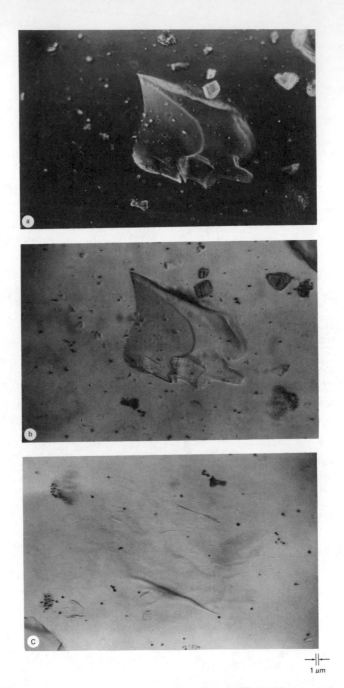

1 μm

FIGURE 9.7 (a) Glass particles, bright contrast, phase.[7] (b) Glass particles, dark contrast, phase.[7] (c) Glass particles, bright field. Same field, same mounting medium (Clarite) as Figure 9.7a and b.[7] Poor contrast.

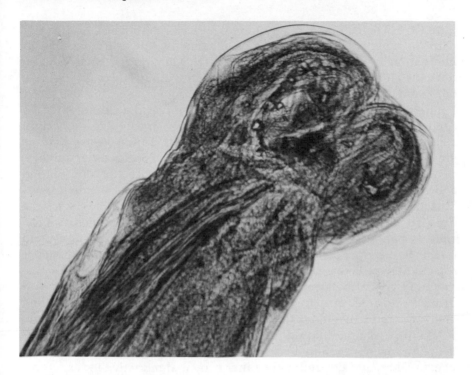

FIGURE 9.8.   Head of unstained nematode worm; dark contrast phase. Electronic flash [7]

annuli are interchanged in the condenser by rotating them in a turret
fitted with labeled click stops.[6] The turret must be centerable so that a
particular annulus may be made conjugate with the corresponding objective's ringed phase plate. The coincidence may be observed by means
of a special compound microscope ("telescope"[4]) in place of the
eyepiece. If your microscope is already fitted with a Bertrand lens
(Chapter 7), use it instead.

9.4.   PHASE-AMPLITUDE CONTRAST IN DETERMINING REFRACTIVE INDEX

Figures 9.7a,b, and c indicate that phase-amplitude contrast may
be helpful in determining refractive index by immersion methods, especially as the matching of refractive indices between specimen and
immersion medium becomes close, as in Figure 9.7c. It is reported that
precision and accuracy in determining refractive index by the Becke test
is advanced an additional decimal point by means of phase-amplitude

contrast.[5] The commercial equipment described above is useful only with isotropic or weakly anisotropic specimens, however, since the illumination is conoscopic. As discussed in Chapter 7, the Becke test depends on the passage of strictly unidirectional rays through the crystal,[6] and the use of conoscopic illumination violates this principle. Figure 9.9 indicates one way out of the difficulty: (a) a clear slot for the condenser, and (b) a groove for dark contrast, or a ridge (for bright contrast) inserted in the objective. Both the slot and the groove or ridge are at right angles to the direction or vibration of the polarized light.

The slot provides the *plane* that contains the direction of vibration and the (perpendicular) direction of propagation of the incident light. This, the plane of vibration, is *parallel* to only one axis of the crystal at a time. In the one view of the crystal as mounted on the microscopical slide, we are interested in only two axes of the crystal; they correspond to the two perpendicular positions of extinction when the crystal is rotated between crossed polars. Therefore, the procedure is to rotate the mounted crystal between crossed polars until one of the two perpendicular positions of extinction is reached. Remove the analyzer and match the refractive index for light vibrating from the polarizer (Figure 9.9a) with a standard liquid (i.e., of known refractive index). After this is done, turn the mounted crystal to the other (perpendicular) position of extinction, and determine the other principal refractive index for this particular view of the crystal. This method is more accurate than the ordinary Becke method because the accessories shown in Figure 9.9 put more contrast into the Becke band.[6]

Unfortunately, such accessories may not be commerically available for the particular microscope being used. Saylor[8] has described how to make a negative phase strip in the laboratory. Hartshorne and Stuart[6] suggest how to make a corresponding substage slot. An alternative

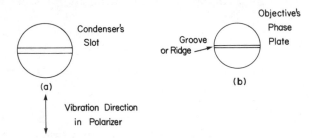

FIGURE 9.9. Special phase-amplitude plates. Substage slot (a) and directional phase plate (b) for determining a specific refractive index of an anisotropic specimen by the Becke test.[6]

method is to use regular phase-amplitude equipment (Figure 9.5), but to place a slot diaphragm perpendicular to the direction of vibration of the polarizer *over* the substage annulus. The result is as before, but with much reduced intensity.[6]

Whereas phase-amplitude contrast is especially applicable in the examination of parts or particles which differ little in refractive index from their natural or designed mounting medium, phase-amplitude contrast is also desirable in the examination of very thin sections. One application is in the sampling of ultrathin sections for transmission-electron microscopy (Chapter 12). Another application is in the deliberate thinning of sections so as to bring them within the prevailing depth of focus and thereby improve the sharpness of the image.[9]

Still the problem of sampling dry ultrathin sections remains: to examine these and other uncovered specimens in air, objectives corrected for use without a cover glass are used, and transmitted phase-contrast illumination is provided by a properly diaphragmed condenser.[9]

9.5. VARIABLE PHASE-AMPLITUDE MICROSCOPY

Microscopists employing fixed phase-amplitude plates sometimes need bright contrast on a darker background, or dark contrast on a brighter background, with increased or decreased contrast within a specimen. Such specimens include unstained cells and tissues, biopsy tissues, ultrathin sections, and any other unalterable system of poor refractive contrast. The Polanret™ (pronounced Po lan' ret) microscope is one of those available for providing continuously variable alteration of phase and amplitude. Polanret™ is coined from *pol*arizing, *an*alyzing, and *ret*arding. As Figure 9.10 indicates, there is a polarizer and analyzer and a $\frac{1}{4}$-wave retardation plate. The diagram also shows a turret of four Polazone™ plates, one for each objective: NA = 0.25, NA = 0.50, NA = 0.66, and NA = 1.25 (oil). Each Polazone™ plate has the conjugate area and the complementary area polarized at right angles to each other. The upper right insert in Figure 9.10 indicates that there are four types of phase plates (A, B, C, and D): solid area, absorbing film, stippled area, and dielectric film. Figure 9.11 is a pair of phase-amplitude photomicrographs taken at optimum selective settings on the Polanret™ microscope.[10,11]

Figure 9.12 is another pair of phase-amplitude photomicrographs: of protoplasmic bridges, *Diospyros discolor*, taken at optimum selective settings on the Polanret™ microscope.[10]

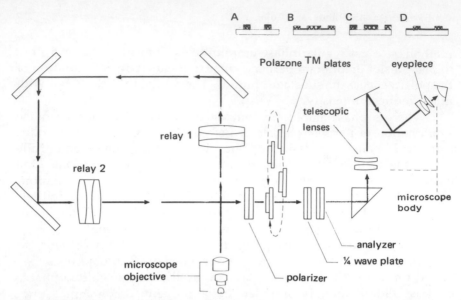

FIGURE 9.10. Diagram of the Polanret™ optical system. Upper right insert: Cross-section diagrams of the four types of diffraction plates. Solid area, absorbing film; cross-hatched area, dielectric film.[10]

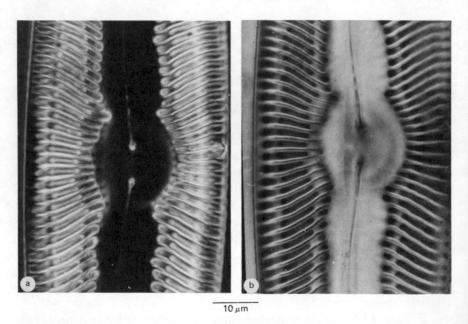

10 μm

FIGURE 9.11. Diatom, *Navicula lyra*, photomicrographed with Polanret™ variable phase-amplitude microscope.[10] (a) Bright contrast plate A, dial setting 0.3; retardation setting 0.25λ. (b) Dark contrast plate B, dial setting 0.2; retardation setting 0.25λ.[10]

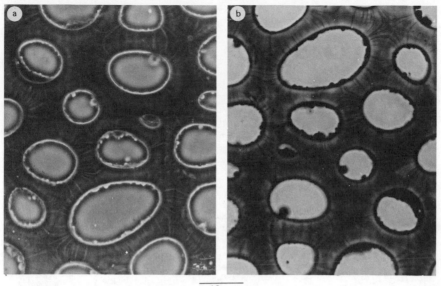

10 μm

FIGURE 9.12. Protoplasmic bridges, *Diospyros discolor,* photomicrographed with Polan-ret™ variable phase-amplitude microscope.[10] (a) Bright contrast plate A, dial setting 0.1; retardation setting 0.5λ. (b) Dark contrast plate B, dial setting 0.55; retardation setting 0.5λ.[10]

9.6. MODULATION-CONTRAST MICROSCOPY

In Chapter 2 there is discussion of the use of oblique illumination to increase resolution and of unidirectional oblique illumination to increase contrast. In our present discussion of the Polanret™ phase-amplitude microscope, the variable crossing of polars[1] is mentioned as a means of modulating the intensity of a beam of light.

The novelty in Hoffman's system[12,13] for modulating contrast lies in the variety of effects obtainable by the use of a sliding, rotatable, slit diaphragm of variable width, all fitted under the condenser. The variable slit is imaged in the back focal plane of the objective, where the modulator is located with its permanently dark, gray, and bright segments. The positioning and adjustment of these components need to be carefully controlled if optimum visibility is to be achieved.[12] Indeed, the components must be adapted and especially fitted to the particular manufacturer's model.[13]

Figure 9.13[13] illustrates the general principles of modulation con-

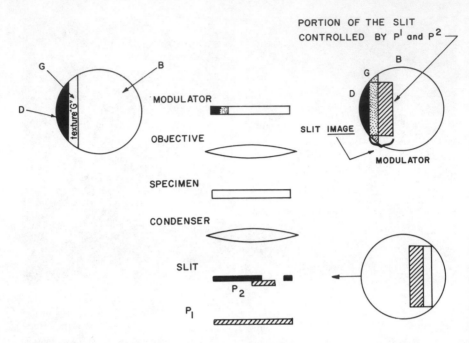

FIGURE 9.13. A diagram showing the components for converting a bright-field microscope to a modulation-contrast microscope. The left plan view shows the modulator regions; dark (D), gray (G), and bright (B). The right plan views shows the slit image correctly registered and superimposed on the modulator. P_1 and P_2 are polars.[13]

trast. P_1 represents a rotatable polarizer of incident light. Behind the polarizer P_1 and in front of the condenser is a sliding slit aperture with a second polar P_2 covering a variable part of the slit. The combination of slit and P_1 is a rotatable slide. The modulator with its dark region (D), gray region (G), and bright region (B) is placed in the back focal plane of the objective.

Centered Köhler's illumination (Chapter 2) is used in the system. The modulator (upper left of Figure 9.13) is viewed using a telescope or eyepiece and a Bertrand lens (Chapter 5). The slit plus polarizer P_2 are slid into place and oriented so that their *image* is superimposed on the modulator, as shown in the upper right of Figure 9.13. The width of the P_2 area (cross-hatched) is controlled by the degree to which the operator slides P_2. The darkness of the P_2 area is controlled by the rotation of the polarizer P_2. Focusing of the slit and P_2 is performed with the condenser. Finally the telescope (or Bertrand lens) is removed, and the eyepiece is replaced (or left in).[13]

Following are a few examples of what can be done with modulated contrast. Figure 9.14 is a fresh human cheek cell, showing the three-dimensional appearance manifested by the shadows alongside the bacteria and the cell's folds.[13] Figure 9.15 shows carcinoma cells from a mouse, comparing (a) modulation contrast with (b) phase contrast. Figure 9.16 shows that modulated contrast may be used on a stained section to give more detail than that manifested by bright field. Figure 9.17 compares photomicrographs taken by modulation contrast (a), phase contrast (b), and bright field (c). The indefinite halo (H) around the phase-contrast image (b) presents difficulty in locating the edge when measuring the diameter, area, or volume. The bright-field image (c) shows a sharp circumference, but with little contrast for micrometry. Micrograph (a), taken by modulated contrast, shows a good sharp edge and minute detail.[13]

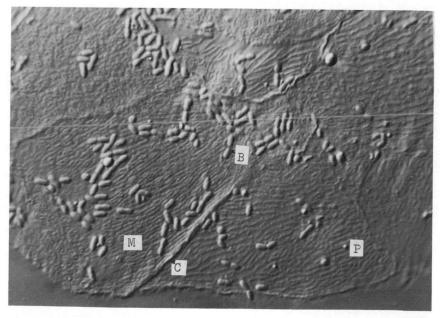

4 µm

FIGURE 9.14. Modulation-contrast image of the surface of a fresh human cheek cell showing the three-dimensional appearance of bacteria (B), membrane folds (M), cell folds (C), and small particles (P). Taken with a 100× Neofluar objective.[13] Courtesy of Robert Hoffman (1977).

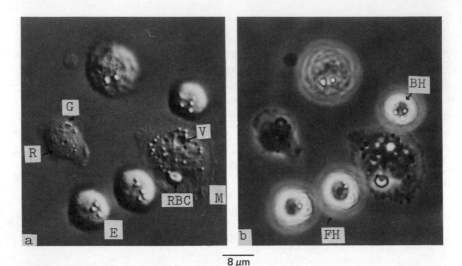

8 μm

FIGURE 9.15. Comparison photomicrographs using modulation contrast (a) and phase contrast (b) to view mouse peritoneal exudate containing Ehrlich carcinoma cells (E) and a relatively flat macrophage (M). Red blood cell (RBC), vacuole (V), and granules (G) are more apparent in the modulation contrast view. Cell R shows multidirectional resolution of the granules. The bright halo (BH) in phase contrast corresponds to position of optical gradients revealed by modulation contrast. The faint halo (FH) in phase contrast is an artifact.[13] Courtesy of M. Padnos and R. Hoffman.

9.7. DISPERSION STAINING

Optical dispersion is the variation of refractive index n with wavelength λ.[1] One way of expressing the dispersion of a particular substance is by the nu value, ν[14]

$$\nu = \frac{n_D - 1}{n_F - n_C}$$

wherein D, C, and F are particular wavelengths λ in the visible spectrum, usually chosen as follows:

$$D = 589 \text{ nm}, \quad C = 656 \text{ nm}, \quad F = 486 \text{ nm}$$

If a specimen is immersed in a fluid of refractive index the same as its own for a single wavelength λ and the dispersion is quite *different* for the two media, either the specimen or the mounting medium will show color contrast, depending upon the position of focus. The phenomenon

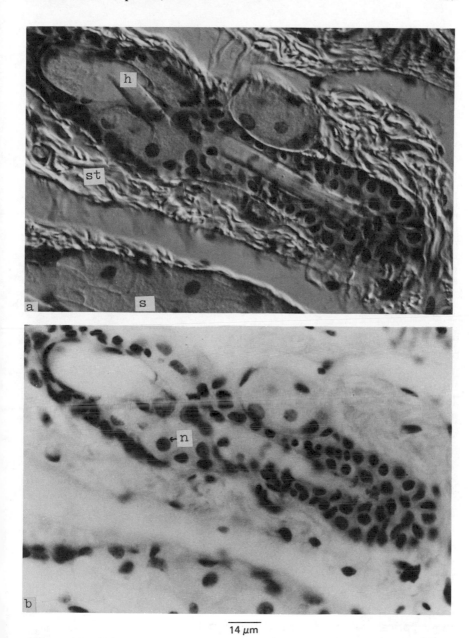

14 μm

FIGURE 9.16. Comparison of photomicrographs using modulation contrast (a) and bright field (b) to view a cross section of mouse skin stained with hematoxalin and eosin.[13] The granular nature of the secretory cells (s), hair shaft (h), and stroma (st) are clearly revealed by modulation contrast. The nucleus (n) is clearly revealed by both systems. Courtesy of Robert Hoffman (1977).

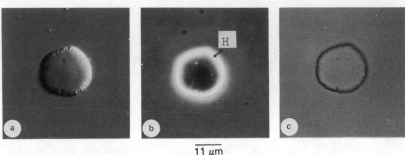

FIGURE 9.17. Photomicrographs comparing the image of a rounded phase object by modulation contrast (a), phase contrast (b), and bright field (c).[13] In (a) note the three-dimensional appearance, image dimension, and edge detail. In (b) the edge detail is only partially visible. The image dimension is indefinite if the bright halo (H) is considered to be an artifact. The halo, however, represents structure and corresponds to optical gradients. The bright-field image (c) is barely visible except for its extent and some edge detail. Courtesy of M. Padnos.

is called the Christiansen effect,[5] and it is often noticed in the determination of refractive index by the Becke method. The Christiansen color may be natural or fortuitous. It can also be produced purposely, and then the process becomes a kind of *optical staining*. Crossman, who advocated *dispersion staining* in 1949,[15] used mixtures of cinnamic aldehyde with butyl carbitol[16] to vary the refractive indices to suit the specimen being optically stained. Various other mixtures are suggested by Hartshorne and Stuart,[17] including a commercially available[18] series of miscible liquids. Each bottle of liquid is labeled with the refractive index for the D line, its temperature coefficient, and its dispersion.

McCrone and his colleagues have published data for the dispersion staining of hundreds of minerals, chemicals, and other materials.[19] In each case, at a given temperature (25°C), values are given for n_F (486 nm), n_D (589 nm), and n_C (656 nm). *It must be clearly understood* that these refractive indices are of *three different liquids* measured for the *D line* at 25°C.[17,20] The liquids are those whose refractive indices *match* that of the specimen at the specific wavelength (color). While McCrone's tables are primarily for determinative purposes, they also give information about optical staining to improve contrast.

Generally speaking, liquids have greater optical dispersion than solids, as shown in Figure 9.18.[14] Therefore, if a particle of a typical solid is microscopically mounted in a typical liquid in which the refractive index of the solid in red light (n_C) is *greater* than that (n_C) for the liquid, and in which the refractive index of the solid in blue light (n_F) is

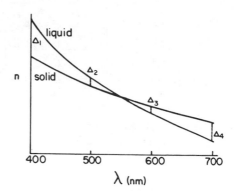

FIGURE 9.18. Dispersion curves.[14]

less than that (n_F) of the liquid, as the objective is focused *upward* the particle will appear red and the liquid will appear blue, as indicated in Figure 9.19.[14] The light (λ_0) of matching refractive index (say $n_{25}{}^D$) will not be refracted but will be transmitted parallel to the optical axis of the microscope. The corresponding analytical dispersion staining "curve" (A) is shown in Figure 9.19. The abscissa is $1/\lambda^2$ instead of λ (wavelength for matching refractive index of liquid and solid) so as to straighten out

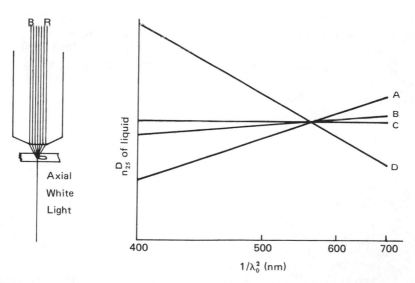

FIGURE 9.19. Refraction occurring at particle liquid interface (left) corresponding to dispersion staining curve A (right); other possible dispersion staining curves are also shown (right).[14]

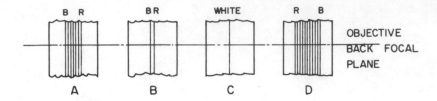

FIGURE 9.20. Objective back focal planes corresponding to dispersion staining curves in Figure 9.19 (right). The specimen is assumed to be that shown in Figure 9.19 (left).[14]

the curves. In "curve" B of Figure 9.19, the difference in dispersion between liquid and solid is much less than in A. In C there is no difference. In D the solid has greater dispersion than the liquid. The corresponding spectra seen in the back aperture of the objective are indicated in Figure 9.20.[14]

9.8. SPECIAL ACCESSORIES

For situations A and B of Figure 9.20, a large annular stop will transmit the central rays Y, whereas a large central stop will shut them off. In situation C of Figure 9.20 (rare), no stop will help the color contrast significantly. In situation D of Figure 9.20 (unusual), the coloration is the reverse of situations A and B.[14] In 1962 Malies designed a turret with two stops which is fitted with the standard objective thread.[21] The housing screws into a standard[22] microscope and carries an objective of NA = 0.25 (Figure 9.21). The turret in the housing has three openings: plain, annular stop (0.2-nm hole), and central stop.[21] These produce the three effects noted above.

FIGURE 9.21. Turret carrying three openings (plain, annular stop, and central stop) over the objective, NA = 0.25.[21] Figure courtesy of Microscope Publications Ltd.

9.9. THE SCHLIEREN MICROSCOPE

The schlieren microscope[23] utilizes stops only 10–12 μm in diameter, and therefore is the most sensitive dispersion staining system.[14] Such small stops cannot be used with objectives of high NA, wherein they would do the most good, because the back focal plane of such objectives is inside the objectives. Instead, the stop is placed at the eye point of an ordinary eyepiece with an auxiliary eyepiece placed above, as shown in Figure 9.22. The original eyepiece now becomes a transfer lens. The new eyepiece should be a 1× telescope. A well-corrected, high-aperture ($f = 1.5$), short-focal-length (50-mm) camera lens is adequate. Its mount should be centerable with the condenser's iris diaphragm. The overall tube length is 200 mm (instead of 160–170 mm).

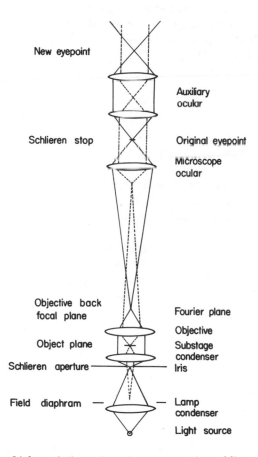

FIGURE 9.22.   Light path through a microscope with a schlieren eyepiece.[23]

Köhler's illumination is used. After the proper schlieren stop is placed at the eye point, the condenser's iris diaphragm is opened as far as possible while maintaining a dark field. A high-NA objective may be used, even an oil-immersion objective.[24]

The use of dispersion staining for determining specific refractive indices of an anisotropic substance requires no monochromator, special illuminators or filters, no match in index with a specific liquid, and no problem with ambiguous movement of the Becke line.[23] It does require determining the dispersion staining curve by noting the matching wavelength (λ_0) for several different liquids in the matching range. The intersection of the best line through $\lambda = 589.3$ nm is n_D. The index n_F is given on the bottle for λ_0 at 486.1 nm, and n_C is given on the bottle for λ_0 at 656.3 nm.[14]

Alternatively, the three λ_0's may be determined by mounting the specimen in a Cargille liquid showing λ_0 near 700 nm at room temperature. Using the temperature coefficient of refractive index for that liquid enables calculation of the index for each λ_0 as the temperature is raised on a hot stage (Chapter 11).

9.10. SUMMARY

If two adjacent particles of a microscopical specimen are not resolved by the physical quality of the optical system (as explained in Chapter 2), then they cannot be distinguished no matter what the magnification. And if they are resolved but do not stand out against the background with sufficient contrast, they still cannot be seen. This second situation arises frequently in the examination of biological tissue, where the watery material differs very little in refractive index from the watery medium in which it is suspended. In such instances the microscopist can gain the necessary contrast by resorting to phase-contrast, phase-amplitude modulation, color-contrast, continuously variable contrast modulation, or dispersion staining.

Since the diffraction image of an object is produced by various degrees of destructive interference and reinforcement of the light waves from all parts of the object included in the angular aperture of the objective, one way to vary the contrast is to vary the paths of light within that cone. Closing down the iris diaphragm of the condenser reduces the prominence of highly scattering parts of the object and thus emphasizes the nonscattering transparent parts. Conversely, inserting a diaphragm with a small opaque disk into the condenser (with wide-open iris diaphragm) will cut off some direct rays and emphasize the wide-angle rays from highly scattering parts. The extreme of this technique is to cut off all direct rays and produce dark-field illumination, so that the opaque or highly scattering particles in the object appear bright against a black background.

A variation of this technique uses multicolored diaphragms (Rheinberg filters) instead of opaque stops. For example, a condenser diaphragm with a red transparent center and a blue annulus around it will produce a red background against which any highly scattering particles will appear as blue dots. *Phase-amplitude* contrast is achieved by a more elaborate arrangement in which a condenser diaphragm with an opaque central stop is complemented by a phase plate, a small diaphragm inserted in the back focal plane of the objective. The phase plate has deposited on it a film of material that delays the light rays passing through it, e.g., by one-quarter of a cycle, so that the delayed rays interfere with the direct rays and enhance the contrast. If the phase-delay film is deposited on the annulus of the phase plate (and the refractive index of the object is greater than that of the medium), the object appears brighter than its background (the mounting medium). If the phase-delay film is deposited on the central portion of the phase plate instead, then the object appears darker than its background. In either case, the interference caused by phase delay reduces intensity, so a compensating reduction of intensity of the oblique rays is achieved by depositing a very thin film of silver or carbon on the annulus film. The combined effect is then phase amplitude contrast.

There are many devices and accessories designed to achieve phase contrast conveniently, all of which involve modification of the condenser and the objective to include the necessary diaphragms. An arrangement that allows quick change from bright-field to dark-field to phase-contrast illumination with parfocal objectives is preferred. All the devices require provision for centering of the complementary diaphragms. Once the chosen equipment is installed, phase-amplitude contrast can be used to improve the accuracy and precision of determining refractive indices.

There is call for continuously variable alteration of phase and amplitude. The Polanret™-type microscope obtains the desired variability by a system of regular polars, a turret of Polazone™ phase-amplitude polar plates, a turret of objectives, and dial settings on a specialized microscope for varying either bright or dark contrast.

The Hoffman system, for modulating contrast, custom-fits a set of only three added units: a polarizer for the incident light, a rotatable holder for a sliding slit with a partial polar to fit under the condenser, and a "modulator" disk with dark, gray, and bright areas to fit in the back focal plane of the objective. The result is directional oblique illumination with modulated contrast between image and background. Such an arrangement provides modulation shadows and sharp outlines without halos, a very three-dimensional effect. Comparison of the modulated contrast image with that of bright field or of phase amplitude is easy.

Disperson staining takes advantage of the Christiansen color effect which is sometimes noticed in the determination of the refractive index for white light by the Becke method (Chapter 6, pages 125 and 126). The Christiansen color is manifested when the variation of refractive index with wavelength in white light is different (usually greater) for the immersion liquid than for the immersed solid. If, for a typical example, the refractive index in red light is *greater* for the solid than for the liquid, and if the index in blue light is *less* for the solid than for the liquid and the microscopical objective is focused *upward,* the solid will appear to be *red* and the liquid *blue*. In recent years, dispersion staining has gained usage for determinative purposes by virtue of the commercial appearance of the required microscopical accessories and of standardized immersion liquids of high dispersive power.

Interferometry in Microscopy

10.1. INTERFERENCE OF TWO WHOLE BEAMS

In this chapter we are concerned essentially with *two whole micro-scopical beams* (rather than individual rays) which are caused deliberately to interfere with each other. The graphic result is a pattern of *interference fringes* analogous to Newton's rings.[1] With incident white light, the fringes are those of Newton's series of color bands more or less superimposed on the pictorial image. Figure 10.1, for example, shows three different micrographs of the same surface area of crystalline grains.[2] All three micrographs were taken on the same simple microscopical interferometer shown schematically in Figure 10.2. Micrograph (a) in Figure 10.1 was taken with practically no tilting angle α to the reference surface (4 in Figure 10.2); hence there was practically no interference. Incidentally, the reference beam was sufficiently out of phase with the specimen's beam to produce *interference contrast*. Thus interference microscopy is related to phase-amplitude contrast (Chapter 9).

Photomicrographs (b) and (c) of Figure 10.1 show definite interference bands. They are contour bands, parallel within grain boundaries but disjointed at the boundaries, because the grain surfaces are at uneven levels due to various etching rates corresponding to crystallographic orientations of the three-dimensional grains. The differences in detectable surface level among grains is slight, varying from a fraction of a wavelength to 10 μm (1958)[1] to 0.5 μm (1970).[3] While interferometry is a delicate kind of vertical micrometry, for purposes of explaining the significance of the interference fringes as contour lines or bands, our model (Figure 10.3) is crudely made for the sake of simplicity. Nevertheless, the stepwise elevation indicated in the side view (a) of Figure 10.3 is suggested by the microsteps on the faces of some crystals which show a plan view similar to that of the model (b).[4]

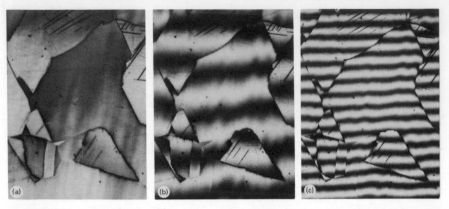

FIGURE 10.1. Three micrographs of the same surface area of crystalline grains, (a) Interference contrast, (b) Broad bands, and (c) Narrow bands, taken on the same interference microscope (Figure 10.2), but with the interfering reference beam at a tilting angle α varying from zero for (a) to greater in (b) and greatest in (c), so that the band width varies from infinity in (a) to narrowest in (c).[2] Taken through Zeiss optics. Courtesy of Carl Zeiss, Inc., New York.

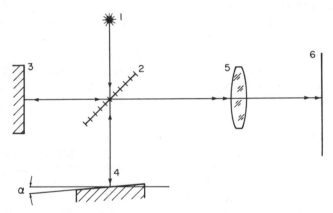

FIGURE 10.2. Schematic diagram of a typical beam-splitting interference system.[2]

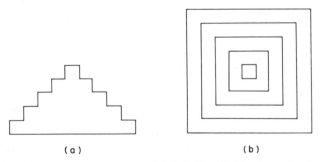

(a)   (b)

FIGURE 10.3. Model of a stepwise pyramidal elevation: (a) side view, showing elevation; (b) plan view, showing contour lines.

10.2. KINDS OF INTERFERENCE MICROSCOPES

10.2.1. Single Microscopes

In Figure 10.2,[2] light from the source (1) is split into two beams by the semitransparent mirror (2). One beam is reflected from the specimen's surface (3). The other beam is reflected from a reference reflector (4) and from the semitransparent reflector (2) into the simple or compound microscope (5). Along the way to the screen (6) (or eye) the reference beam interferes with the beam from the specimen (3), producing interference images such as are shown in Figure 10.1.[2]

While the single-lens system of Figure 10.2 is that of a simple microscope with the semitransparent reflector (2) in front of the lens system, a prototype interference microscope is based on a microscope fitted with vertical illuminator (Chapter 4) and an optically flat glass plate in close contact with the surface of the specimen. If it is a metallic specimen, its surface is too reflecting when compared with the surface of the glass plate. Therefore, the glass plate is given a semireflecting coating of commensurate reflectance, employing silver or other metal. With monochromatic light, adjacent fringes correspond to differences in depth of half a wavelength, so that even shallower depths may be measured. This principle has been modified by Mirau, who converted the "glass plate" into a beam splitter so that one beam is directed to a separate metal-coated reference surface while the second beam is reflected from the surface under examination, as shown in Figure 10.4.[5] The Watson manufacturers have converted the system into a *unit objective* by bringing the illumination into the side of the objective instead of the side of the microscope's tube. Therefore, the "semireflecting surface" of Figure 10.4, or the equivalent, must be removed if the microscope is fitted with a vertical illuminator. As shown in Figure 10.5, the Watson interference objective will fit on almost any microscope equipped with a standard thread.[6] The light from a sodium lamp (no filter) or mercury lamp (plus D filter) enters the collimating condenser C on the side of the objective. The beam splitter E sends some light down to the specimen (and then through the objective system O). The beam splitter E sends its second beam to the reference mirror L. The reference beam from L is aimed back to the beam splitter E by adjusting screws G, H, and K, so that the reference beam is reflected into the objective system O. There the reference beam interferes with the beam from the object to produce tell-tale fringes depicting the specimen's surface. Shutter F shuts off the interfering beam when the specimen is to be observed without the interference fringes.[5]

The above-mentioned single beam-splitting interference micro-

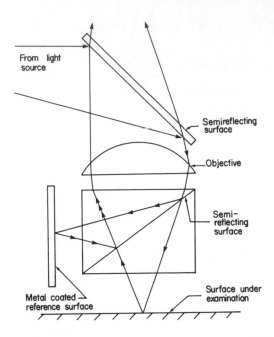

FIGURE 10.4. Mirau interference objective principle.[5] Courtesy of Microscope Publications, Ltd.

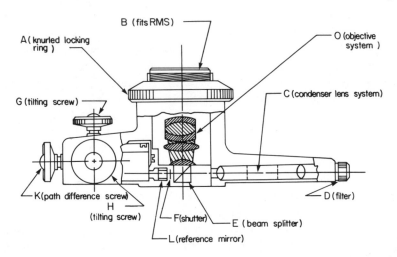

FIGURE 10.5. Watson interference objective. (Drawing adapted from illustration courtesy The Watson Microscopy Division of M. E. L. Equipment Company Ltd.)[5] Courtesy of Microscope Publications, Ltd.

scopes (Figures 10.2, 10.4, and 10.5) have been utilized in examining metallic surfaces and edges (such as razor blades) and thin films (such as lacquers and electron-microscopical substrata). Thin biological specimens can be studied in similar fashion by mounting them between half-metallized optical flats. In this case, the internal structural features of the specimen govern the local resistance to cutting, and so the features are reflected in the surface topography of the thin section.[1]

The interference microscopes as illustrated by Figures 10.2, 10.4, and 10.5 have this in common: The specimen surface and the reference surface are separated by a few centimeters, and so the two wavefronts may be separated by as much. However, another way of separating the two wavefronts is to employ a doubly refracting substance such as calcite (Chapter 5). Then the separation of the two wavefronts is reduced to as little as the resolving power of the light microscope ($\approx$0.2 μm; see Chapter 2). Such fine beam splitting is called *differential* interference contrast, or differential separation of the wavefronts.[2] Figure 10.6 is a schematic diagram of differential beam splitting by *transmitted* light, and Figure 10.7 shows beam splitting by *reflected* light.[2]

Both microscopes (Figures 10.6 and 10.7) feature the Wollaston prism as modified by Nomarski.[2] This prism receives polarized light and splits it into two wavefronts that are separated by only 1 μm

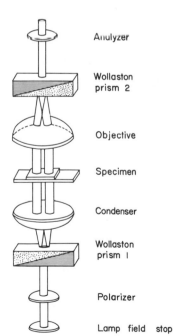

Analyzer

Wollaston prism 2

Objective

Specimen

Condenser

Wollaston prism I

Polarizer

Lamp field stop

FIGURE 10.6. Schematic diagram of a two-beam interference microscope for differential interference-contrast in transmitted light. (Separation between the two wavefronts is exaggerated.)[2]

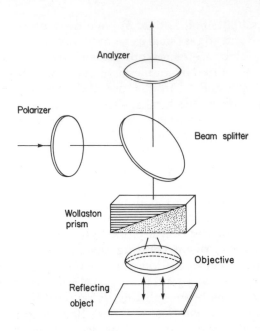

FIGURE 10.7. Schematic diagram of a reflected-light system for differential interference-contrast microscopy.[2] (Separation between the two wavefronts is exaggerated.)

(exaggerated in Figures 10.6 and 10.7). The Zeiss–Nomarski equipment is designed to complement the Zeiss phase-amplitude contrast equipment. This interchangeability is in itself an advantage. At the same time, it emphasizes that the chief purpose of this type of interference equipment is to gain contrast. The contrast is gained without loss in resolution or optical sectionability; at the same time there are no troublesome halos and no loss in path length. The brilliance of colors is especially striking by reflected light.[2] The Nomarski method is also recommended for the examination of polymer crystals, which are otherwise low in contrast[4] because they differ so little from their surroundings in refractive index.

The Baker interference microscope incorporates a continuously variable phase difference between images in and out of focus, as produced by doubly refractive rotatable plates. The image is in color contrast which shows very small differences in thickness and refractive index. Like the Nomarski microscope, the American Optical Baker interference microscope is recommended for the contrast it lends to specimens low in contrast[4] (Figure 10.8).

10.2.2. Double Microscopes

The double interference-contrast microscope, devised by Linnik, is built like Michelson's interferometer.[1] The illuminating beam is split and sent into *two* microscopes: one for the specimen and the duplicate for the reference beam. In the Leitz interference microscope developed by Mach and Zehnder the two beams are separated by 62 mm.[7] The isolation of the two beams removes all concern about the identity of the specimen area and that of the reference area. An anisotropic specimen will produce a fringe shift of a particular magnitude and direction. If the same specimen is placed under the *other* microscope, the fringe shift is equal but *opposite*.[7]

An important use of this type of interference microscope is in the determination of refractive indices.[7,8] Since most specimens are anisotropic and therefore have two or more specific indices, the instrument uses incident light which is polarized. Part of the polarized light passes through the specimen, which is mounted in a liquid of known refractive index and is oriented at one of the two positions of extinction. The other part of the polarized light goes through only the standard mountant. The two beams are then combined and the resultant retardations appear as amplitude differences in the image. These are measured by means of a built-in compensator such as the Sénarmont type. If the thickness at

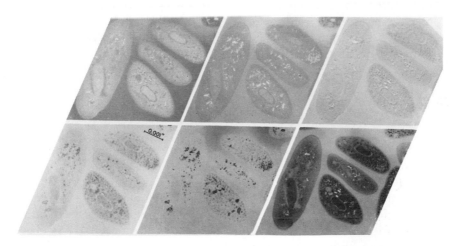

FIGURE 10.8. Taken with the A-O Baker interference microscope. Contrast can be changed to show desired detail, as in these photomicrographs of unstained protozoa—part of a series used in the *Scientific American* and on educational TV. Courtesy of American Optical Co. and R. W. Richards.

one given point in the specimen is known, the refractive index can be calculated. Or, the specimen may be mounted in a second standard liquid and the second retardation can be observed as before. Then it becomes possible to calculate the refractive index without knowing the thickness. This method gives a more accurate value for the refractive index than the Becke method. Besides, interferometry gives the *average* value through the cross section of the specimen, while the Becke method gives the refractive index of the superficial layer. Therefore, if the specimen is a coated (e.g., weathered) crystal or a fiber with a characteristic skin, the values by the two different methods will be different and thereby determinative.

10.3. APPLICATIONS TO HIGHLY BIREFRINGENT SPECIMENS

If the birefringence of a man-made fiber is low, the fringe shifts for $n_{\parallel}$ and $n_{\perp}$ are easily determined. Then if the thickness of the fiber is known, the actual values for $n_{\parallel}$ and $n_{\perp}$ can also be determined (Chapter 6). However, when the fringe shift is more than three orders the thickness of the sample changes so rapidly at the edge of the fiber that it becomes difficult to assign an exact order number to a fringe. To solve such a problem, Scott[7] mounted highly birefringent fibers in a liquid of refractive index near $n_{\parallel}$ or $n_{\perp}$. Using *un*polarized light he obtained the Moiré effect of superimposing the interference pattern for $n_{\parallel}$ in the fiber upon the interference pattern for $n_{\perp}$ in the fiber. The resulting dark regions correspond to places where X orders of interference at $n_{\parallel}$ overlap Y orders of interference at $n_{\perp}$, X and Y being whole numbers.

10.4. SUMMARY

Interference microscopy provides a way of gaining contrast by superimposing the image of a specimen and interference rings or bands of controllable width and density. This is accomplished by splitting the illuminating radiation into two separate beams, putting one beam through the sample, introducing a controlled delay or phase difference into the reference beam, and then combining the two beams so that they interfere with each other and produce the rings. If white light is used, the rings are like the familiar Newton's rings in color. When monochromatic light from a sodium lamp or a filtered mercury arc is used, the image is sharper and more easily interpreted. As a further advantage, interference microscopy provides clear perception of very small changes of depth or elevation in the sample, and such changes can be measured by counting the successive interference rings. Still further refinement of the method, using polarized light, allows precise calculation of the refractive indices of birefringent fibers and crystals.

An interference microscope may be quite simple or extremely complex, depending upon what is required of it. The simplest arrangement splits the illuminating beam by means of a lightly silvered semireflecting glass plate placed at an angle of 45°, then reflects one beam from the sample while the other beam is reflected from a reference surface which may be tilted through small angles to introduce a phase difference. The two beams are then recombined within the optical system of a simple or compound microscope to produce the interference rings or bands in the image. Alternatively, the semireflector and its associated illuminating system can be converted into an objective designed for vertical illumination of opaque objects. Thin transparent objects such as biological specimens can be mounted between half-silvered optical flats.

Another way of splitting the illuminating beam is to use a prism made of a doubly refracting crystal such as calcite (the Wollaston or Nomarski prism). The Baker system uses doubly refractive rotatable plates to achieve a continuously variable phase difference. It is especially useful for examining thin crystals or thin sections of minerals.

The greatest flexibility and capability in an interference microscope is achieved by using two separate compound microscopes, one for the object beam and the other for the reference beam. The Leitz interference microscope is an advanced example: it splits the illuminating beam into two parallel beams by means of a prism, sends one beam through the sample via a fully equipped compound microscope, sends the other beam through a duplicate optical system fitted with tiltable-plate phase-delay devices and a wedge compensator, and then combines the two beams in a duplicate prism before sending them through the eyepiece. Both optical systems are fitted with polarizers, analyzers, and all appropriate diaphragms and stops. This arrangement allows the specimen to be moved about at will, and keeps the specimen area separate from the reference area. Furthermore, the specimen can be shifted from one stage to the other, so any fringe shift of a particular magnitude or direction caused by the sample will be reversed when the sample is shifted to the other stage. Thus all confusion is removed from the interpretation.

When an anisotropic sample is mounted in a liquid of known refractive index and is oriented between crossed polars of an interference microscope, the resultant retardations appear as rings in the image. These rings can be counted and measured by means of a built-in compensator. If the thickness of the specimen at any given point is known, the refractive index can be calculated. When two successive mounting liquids are used, the index can be calculated without knowing the thickness. The method gives an average refractive index for the entire thickness of the sample, rather than an index for the surface only, as in the Abbé method. Even highly birefringent fibers, difficult to deal with by any other method, can be measured by interference microscopy by observing the number and direction of interference bands, as explained in the text.

Microscopical Stages

11.1. INTRODUCTION

Practically all of microscopy is staged in some way and to some degree. In its simplest state, a microscopical *stage* is a device used to hold the specimen in a desired position in the optical path.[1] On stereoscopic, biological, and metallographic microscopes the fundamental stage is usually rectangular, flat, and drilled to take clamps for holding the microscopical slide or the specimen itself. Some rectangular stages are made with built-on *mechanical stages* (Figure 11.1).[2] Otherwise, there are usually tapped holes to take the fastening screws of a mechanical stage[2-3] for moving the specimen in X and Y directions. Graduations plus their vernier scale on *rack-and-pinion scales* usually allow for measurements in tenths of a millimeter.[3] Some specialized mechanical stages[2-6] operate with *micrometer screws*, which with vernier measure to within hundredths of a millimeter (Figure 11.2). Micrometer screws may be integrated into a single mechanical stage so that in a single traverse of a composite specimen the paths over separate constituents may be measured separately (Figure 11.3).[2] This mechanical idea for areal analysis has been automated by means of motors and counters.

All stages should be waterproof and oilproof. They should also be reasonably resistant to solvents, chemicals, and heat. The plane of the main stage must be exactly perpendicular to the axis of the microscope; otherwise the optical quality is set back hundreds of years. If the microscope is ever dropped or the stage is struck in any way, the microscope should be carefully examined, and if damaged, repaired. The central opening should be at least 25 mm in diameter so that the condenser may be raised level with the upper surface of the stage.[7]

Metallographic and other microscopes for thick specimens should provide for movement in the third (Z) direction (Figures 11.3 and 11.4).

FIGURE 11.1. This biological microscope has a built-on, graduated mechanical stage.[2]

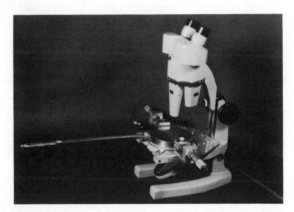

FIGURE 11.2. Film stretcher equipped with graduated micrometer screws and Köfler hot stage.[4]

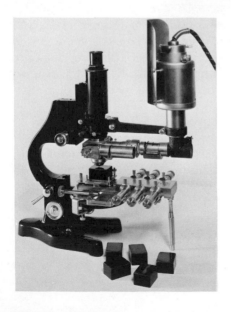

FIGURE 11.3. Integrating stage (Leitz). Its six micrometer screws permit the quantitative approximation of six constituents or groups of constituents in a single traverse of the specimen.[2]

FIGURE 11.4. Rectangular briquet fitted in a mechanical stage (Leitz) on a rotatable stage (Leitz).[2]

Polarizing microscopes should be provided with a *rotatable stage, graduated* in degrees (and a vernier in tenths) at the circumference (Figure 11.4).[2] Centering screws should be provided unless the objectives and condenser are centerable. Ball bearings insure against wear and keep the stage from wobbling when turned. A rotatable stage should be large enough so that the central 40 mm of a 75-mm slide can be explored and rotated without having the slide hang over the edge of the stage. If the rotatable stage is easily removable, a hot stage or other accessory can be inserted in its place. A removable rotatable stage is also easily lubricated and otherwise serviced.[7] A locking screw or other device will prevent rotation whenever a small attached mechanical stage is used (Figure 11.4).[2]

The *glide stage* is relatively new (Figure 11.5).[3] It is designed to speed up the scanning or searching of an entire large specimen, such as a biological or crystalline culture in a Petri dish. Glide stages should provide ease of motion on the glide plane and rotational bearing, and yet be resistant to slight unintentional pressures.

There are many special stages and holders.[3,7] Some are designed to level a polished section so as to be perpendicular to vertical illumination. Other stages and holders are for special objects, such as a chuck for bullets.[3] This and other chucks may be used for a variety of purposes. Figure 11.6, for example, shows a chuck being used to hold a "dental" drill for removing a microsample from an ore fraction.[2] The chuck in this case is on a separate stand (coarse micromanipulator) because the

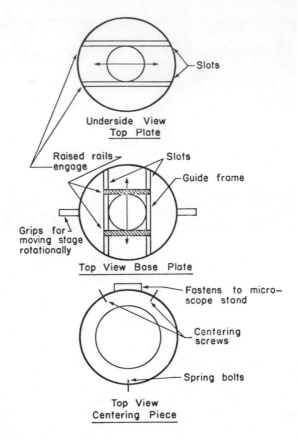

FIGURE 11.5.   Glide stage principles. (After photographs courtesy of Carl Zeiss.)[3] Courtesy of Microscope Publications, Ltd.

FIGURE 11.6.   Dental drill and micromanipulator. The chuck of the dental drill is held in a special clamp attached to a Leitz-type micromanipulator.

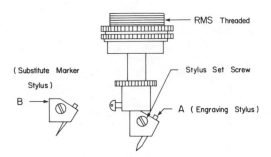

FIGURE 11.7. Sheaff microobject marker. (After illustration courtesy of Arthur H. Thomas Company.)[3] Courtesy of Microscope Publications, Ltd

rotating cable and its sheath are too stiff to be mounted directly on the microscopical stage.

Another important use for the mechanical stage is to locate and record a certain position on a specimen by means of numerical coordinates x and y. To be returned to a position, the specimen must be put back into the mechanical stage in the original orientation of front vs. back and left vs. right. A rectangular specimen such as a microscopical slide or a rectangular briquet (Figure 11.3), properly labeled, offers no problem of orientation. Circular specimens need an index such as a notch to be engaged by a lug. The return to a location is initiated with a low-power objective.[2,3]

Another way of returning to a location in a specimen is to mark the spot with a qualitative marker (Figure 11.7[3]) or a quantitative hardness indenter (Figure 11.8[2]) that fits into the standard thread.[8] The Bier-

FIGURE 11.8. Vickers-type impression hardness tester (Eberbach). The Vickers-shaped diamond point is attached to a spring in a holder fitted with Society objective thread. A standardized compression of the spring, by use of the microscope's fine-focusing adjustment, is indicated by a signal light on the panel of the box shown in the picture.[2]

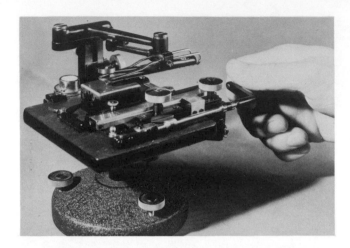

FIGURE 11.9. Bierbaum scratch-hardness tester. The tester fits a square-type micro-scope stage, or the stand shown above. The specimen is moved under the diamond point by means of a micrometer screw. The degree of hardness is indicated by the width of scratch as measured microscopically.[2]

baum scratch-hardness tester[9] can be used to mark a microscopic place. As shown in Figure 11.9,[2] the Bierbaum tester fits a square stage such as that of the stereomicroscope. The test-scratch can be made long enough to be easily located and followed.

11.2. MICROMANIPULATORS

Various operations require mechanical aid in controlling the X, Y, and Z directions of moving a microtool. Coarse movements may be provided by ordinary rack-and-pinion assemblies, as shown in Figure 11.6. The vertical mechanism is much like that of a simple microscopical stand on which the X–Y movements could be added with an ordinary mechanical stage. Somewhat finer adjustments are provided by micrometer threads of ordinary screw-micrometers.[3-5]

Useful magnifications in the range of 300–600× require finer movements in the X and Y directions. The Chambers system is based on one to four units, two of which are presented in Figure 11.10.[3] The X and Y movements in each unit are provided by modified micrometer screws with control knobs which are attached to flexible shafts. The shafts also absorb vibrations from the operator's hands. The clamps for holding the microtools are mounted on vertical rods for leeway in adjusting to the Z direction. Up to four units may be clustered around a

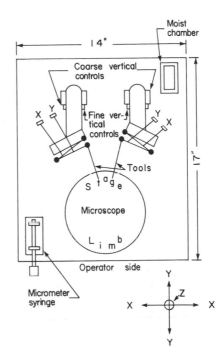

FIGURE 11.10. Typical operational layout, Chambers micromanipulators. (After photographic illustration courtesy Brinkman Instruments, Inc.)[3] Courtesy of Microscope Publications, Ltd.

microscope. Figure 11.10 shows two mounted in a moist chamber for biological purposes.

The Emerson micromanipulator works on a different principle. A single vertical lever (joystick) controls an omnidirectional movement in the plane of the microscopical field. The measurement of the joystick is in the same direction as the apparent direction of motion. The extent of travel can be varied from about 0.025 mm to about 3.2 mm for about 90 mm of travel by the hand. A microworm gear controls movement in a horizontal arc of 20°. The mechanical stage can be tilted through about 10°. The tool holder is a tubular chuck which allows attachment of tubing for a gas or for electrical connections through a hollow needle.[3]

11.3. HEATABLE STAGES

Heatable ("hot") stages are a necessity in the direct microscopical observation of thermal transitions above room temperatures. The maximum temperature to be reached is set by the materials of construction in ordinary light microscopes: about 350°C. Stages heatable to such temperatures are adequate for studying most organic crystalline phases,[10,11] fibers,[12] many other polymeric materials, and many inorganic crystalline transitions.[7] Although most metals require higher temperatures for significant thermal studies, there are parallel physical–chemical transformations among organic systems that are very instructive in the fields of metallography[13] and ceramography. Heatable stages within the range up to 350°C may be divided into two categories: those with long working distances and those with short ones.[11] Usually, the working distance refers to the distance between the specimen *in* the stage and the outside lens surface of the objective which has a high-enough NA to obtain an interference figure.[7] In Chapter 7 we determined that the ideal NA for interference figures is 0.85 because tables of reference between measurements on interference figures and other optical properties are based on an NA of 0.85.[14] This value of 0.85 is not absolute, since there are special objectives of NA $\approx$ 0.85 with longer than ordinary working distance, such as those objectives intended for use with universal stages (see Chapter 7).

11.3.1. Hot Stages with Long Working Distances

The Kofler hot stage is classical[10,11,15] and is made in Austria by Reichert. The American version was designed by H. Alber for the A. H. Thomas Company. It is very useful on stereomicroscopes with long working distances,[4,5] on chemical microscopes,[7,10,15] and wherever interference figures are either not needed or cannot be observed under the circumstances. As shown in Figure 11.11,[16] the Thomas–Kofler hot stage is housed in a metal cylinder A, fitted with metal rim B and glass cover C. The heating coil HE (Figure 11.12) is electrically supplied and regulated by its variable transformer. The stage takes a half-slide (25 × 38 mm). The half-slide can be placed within fork F of Figures 11.11 and 11.12 so that the specimen may be moved from outside the stage. A glass baffle D (Figure 11.11) may be used to ensure more uniform heating. Two thermometers are supplied (30 to 230°C and 60 to 350°C), and each is precalibrated while in its protective metal sheath M. There are many accessories, each with its own heat capacity, so the thermometer to be used with the particular assemblage of accessories should be checked with the appropriate reference melting point standards, CR (Figure 11.11).[16] More precise measurement of the temperature of the specimen may be performed by placing a fine-wire thermocouple *on the specimen* just beyond the microscopical field of view. The other end of the thermocouple is kept in ice + water, bubbled with air to keep its temperature constant, in a Dewar flask.[5]

As seen in Figure 11.11, the Thomas–Kofler stage has its own condenser R. Sometimes, as with a stereoscopic microscope,[4,5] this lens should be removed by taking out the screws RS. The lens is also re-

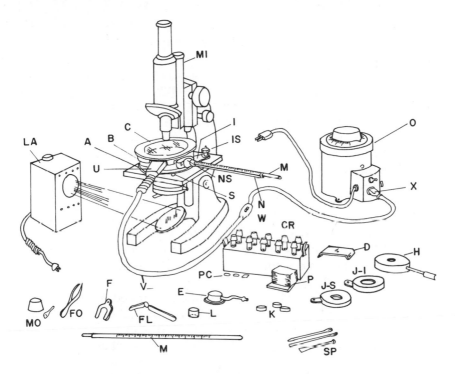

FIGURE 11.11.   Accessories for the Kofler hot stage.[16]

moved when dark-field illumination is intended with a polarizing microscope. For dark field a Leitz objective (UM-4, NA = 0.20) of long working distance is used.[17] The top lens of the regular substage condenser is removed and an adequate central stop (Figure 9.4) is placed in the slot under the condenser. The UM-4 objective has an iris diaphragm which is closed until dark field is obtained. With this setup of the Kofler

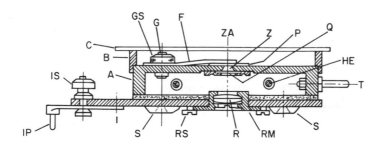

FIGURE 11.12.   Cross section of Kofler microscopical hot stage.[16]

hot stage and dark-field illumination, the growth or decline of crystalline units such as spherulites with variation of temperature has been reported.[17]

The Leitz Company[18] is producing a hot/cold stage for temperatures from 360°C down to −20°C, which allows the use of long working objectives (UM), originally designed for the universal stage. This provision for numerical apertures up to 0.6 permits the observation of useful interference figures.[11]

The Mettler[19] hot stages FP2 and FP52 feature two heating coils, one above the specimen and one underneath, thus practically eliminating the vertical thermal gradient of a single coil. A second advantage is the use of a small fan (P in Figure 11.13[11]) used to circulate the air inside the stage. In the older stage, FP2, the air can be cooled by introducing a cold gas through a fitting underneath the fan. In the newer model, FP52, the cold gas can be introduced directly through a tube in the housing of the stage. Otherwise, the two stages are alike and will be described together.

As shown in Figure 11.13, the specimen is moved by control knobs K. The specimen is thermally protected by a mirrored housing O and by a current of air from fan P. The window Q is also a heat filter which together with O and P protects the microscope as well as the specimen. The temperature, 20 to 300°C, is measured by a platinum resistance thermometer in fixed position. The automatic alternative rates of heating, 0.2°, 2°, or 10° per minute, can be preselected by push buttons, but can be overridden by another push button. The temperature is continuously recorded digitally to within 0.1°C at the top of the panel, while significant events (melting or other transition points) may be recorded by pressing the appropriate button. A special button arrests the temperature and maintains it as long as the button remains depressed.[11] Instructions and correction factors have been published.[20]

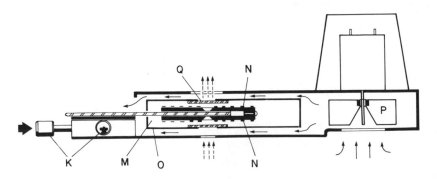

FIGURE 11.13. Mettler FP2 stage: side section (Mettler Instrument AG).[11]

There is an addition to the Mettler FP52 heatable stage: the *photoelectric sensor* which is employed in *thermal depolarization analysis* (TDA) (not differential thermal analysis, DTA).[11] The photoelectric sensor records the changes in light transmission which accompany phase transitions of a specimen heated or cooled between crossed polars[11] (preferably synchronously rotating crossed polars). The most striking transition is melting to the point where the whole field becomes dark.[21] One eyepiece of the binocular microscope is replaced by the photoelectric sensor which signals the changes in light transmission.[11] The technique has been applied to the study of thermal changes in polymers and mesomorphic systems.[21]

The mutual problem with all heatable stages having long working distances is that ordinary objectives with NA sufficiently high for observing interference figures cannot be focused close enough to the specimen. The use of long working objectives, such as those designed for universal stages,[18] has been mentioned as a solution to the problem. Another more general solution which includes the use of ordinary objectives is to employ a stage with short working distance.

11.3.2. Hot Stages with Short Working Distances

Hot stages heated up to 150°C can be tolerated by ordinary objectives of nominal 4-mm focal length (NA = 0.85).[11] The criterion of toleration is that their ability to portray interference figures is not impaired by continual use with the hot stage. Then the problem is to have the working distance of the stage as short as that of a 4-mm objective. Hartshorne designed two thin heatable stages: a circular one to fit onto a polarizing microscope with the conventional circular stage, and a rectangular stage to fasten directly onto a microscope with the stationary rectangular stage.

The circular thin hot stage consists of a single brass disk about 10 cm in diameter and about 4 mm thick. The disk is recessed about 2 mm, leaving a rim for peripheral screws and for a central ring to contain the axial hole. The recess in the disk is to hold the nichrome heating ribbon threaded through holes in a sheet of mica. This heating unit is insulated from the disk by a whole sheet of mica. The axial hole of the disk is tapered to the shape of the cone of light from the condenser (about 4 mm in upper diameter and 8 mm in lower diameter). The hot junction of the thermocouple is made by securing the copper and constantan wires separately to the central boss by means of very small countersunk screws. A third mica sheet covers the stage and is fastened by small screws. There are four knurled hand nuts by which the whole stage is attached to the main rotatable stage of the polarizing microscope. Two knurled nuts are drilled to take stage clips. Heating is controlled by a variable transformer. The resulting voltage is read on a millivoltmeter with a scale that has been calibrated with the known melting temperatures of pure crystalline standards that have been observed to be in equilibrium with their melts at the corresponding voltages.[11] The cold junction of the thermocouple is immersed in ice water bubbled with air in a vacuum flask.[5]

The circular Hartshorne hot stage has two major limitations, both of which arise from the thin design of its heating element in the metallic shell which sits on the microscope's stage. One limitation is that the metallic shell raises the specimen above the substage condenser, which has to furnish conoscopic illumination to obtain an interference figure. Perhaps the stop on the condenser's substage can be adjusted to raise the condenser sufficiently; if not, perhaps an auxiliary lens or different condenser can be inserted. The second limitation to the circular design is that even if the round stage is not heated over 150°C, it may be hot enough to damage the microscope's stage or polarizer, or some other part. A solution may be to substitute a thermal insulator such as an asbestos composite for the metallic shell. (If a thermal insulator is merely stuffed under the metallic stage, the specimen is raised even farther above the condenser.[11]) In any case, the polarizer should be of a polarizing plastic such as Polaroid® rather than of calcite because the former is much less expensive to replace. If the plastic sheet needs to be replaced, the replacement should be of a heat-resistant variety such as the Polaroid Corporation makes.

If the microscopist wishes to go slightly higher than 150°C, he may select Hartshorne's rectangular design because the heating element hangs under (rather than sitting on) the metallic shell. Of course the rectangular model is not rotatable. The stationary aspect means that the electrical leads are also stationary and do not twist back and forth, as with the rotating stage. A point to be considered is that polarizing microscopes with synchronously rotatable polars are no longer produced commercially. If such a microscope is available, the rectangular heatable stage shown in Figure 11.14 is appropriate.[11] The metallic shell of the rectangular stage is made of thin copper sheet (not more than 0.8 mm), rolled to a slight concavity upwards [Figure 11.14a, (i)] so that mounted on the microscope's stage it will straighten out when pulled down at the corners. A thin sheet of mica [Figure 11.14a, (ii)] has a central hole in which a circular cover glass rests on a thicker sheet of mica [Figure 11.14a, (iii)], with an extension carrying the binding posts for the leads from the heating coil and for the thermocouple. This sheet [Figure 11.14a, (iii)] has a central hole about 6 mm in diameter for illumination. The heating coil is of nichrome wire or ribbon that con-

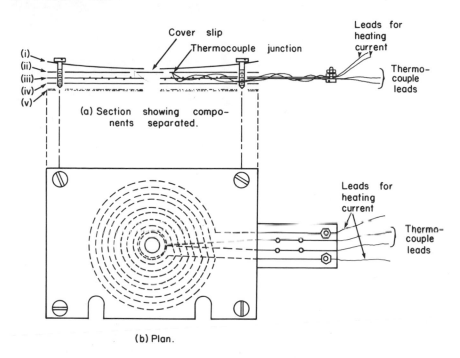

FIGURE 11.14. Simple thin hot stage. (After Hartshorne and Stuart.)[11] Courtesy of Microscope Publications, Ltd.

ducts 1 or 2 amp at about 36 V. The sheet also carries safely and carefully a copper–constantan thermocouple junction as near to the center of the stage as possible. Another thin sheet of mica [Figure 11.14a, (iv)] and one of asbestos paper [Figure 11.14a, (v)] serve as thermal insulators. They carry holes about 6 mm in diameter for low powers and 3 mm for high powers.[11]

11.3.3. Hot-Wire Stage

A hot wire can heat a microscopical specimen directly, between cover glass and slide. The result is a temperature gradient which can be very useful in producing two or more phases of a system in a single microscopical field. For example, if we stretch a wire in a melt that freezes at a temperature within ordinary hot-stage range (between ≈300°C and room temperature), and if we remelt the solid by flowing an electric current through the wire, we may obtain two or more phases on both sides of the wire, as shown in Figure 11.15.[11] Jones[22] uses No. 24 gauge platinum wire about 25 mm long, held between two clips. To

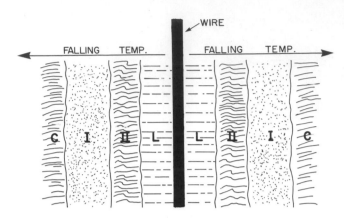

FIGURE 11.15. Diagrammatic representation of field shown by a compound having two mesophases under a hot wire. L, liquid; II, mesophase stable at the higher temperatures; I, mesophase stable at the lower temperatures; C, crystalline phase.[11] Courtesy of Microscope Publications, Ltd.

keep the heating wire straight each clip is soldered to a stiff copper wire held firmly in place. The other ends of the two copper wires are soldered to flexible wires leading to a transformer. The slide and cover glass containing the specimen rest on an aluminum ring to protect the condenser from the hot slide.[11,22]

Hartshorne has elaborated on the hot-wire principle.[23] His wire stage fits onto his polarizing microscope with its rectangular stage. The polars rotate together instead of having a rotatable stage. As shown in Figure 11.16, the nichrome heating wire is stretched on a rectangular block A (6 mm thick) of a hard asbestos composite carrying a central hole 20 mm in diameter. Two brass platelets C,C' are bolted to A and contain binding posts and finger nuts D,D', and G. The nichrome wire E, even when hot, is kept taut by a strong spring hook F. The posts H and H' are of such height that the wire E just misses touching the cover glass over the specimen on slide J. Cover glasses and slides are selected within narrow limits of thickness so that the level of wire E does not need to be readjusted. Clip L holds the slide J tightly against cardboard strips K and K', which thermally insulate the slide from asbestos composite base A. The temperature gradient can be increased by means of a brass or copper cooling plate M, shown in Figure 11.17. Soldered underneath the plate M is a brass shim N which is of the same thickness as the cover slip. The wire E is heated by about 6 V from a variable transformer.

Jones[22] determined melting points by means of a tiny thermocouple. Hartshorne[23] calibrated a linear eyepiece scale oriented at

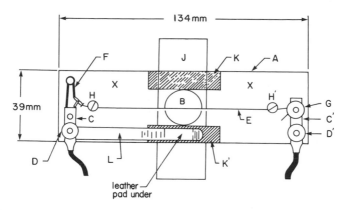

FIGURE 11.16. Hot-wire stage in plan.[23] Courtesy of Microscope Publications, Ltd.

right angles to the heated wire. He used a compound having at least three known transition points[7,11,13] and measured the distance on the eyepiece scale between phase boundaries. (The wire's thermal gradient is not linear and so at least three phases are needed for the calibration of the eyepiece.) The standard and unknown specimen are mounted under separate strips of cover glass, as indicated in Figure 11.18. The strips are melted separately under their separated cover glasses, and any excess solidified substance is cleaned off with a razor blade. After the strips are mounted at right angles to the hot wire, they are heated and observed.[23]

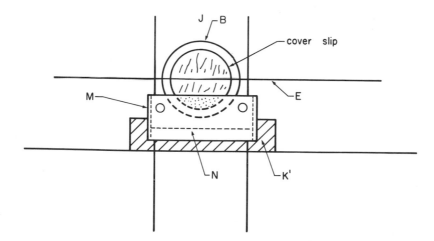

FIGURE 11.17. Cooling plate used to increase temperature gradient. (Labeling common to Figure 11.16 has the same significance.)[23] Courtesy of Microscope Publications, Ltd.

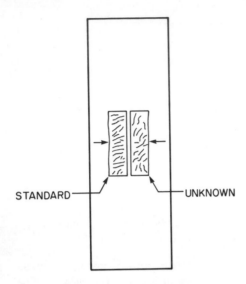

STANDARD—   —UNKNOWN

FIGURE 11.18. Comparison preparation for determination of transition
temperatures. Courtesy of Microscope
Publications, Ltd.

11.4.   VERY HOT STAGE

Temperatures up to 1000°C may be obtained on a hot stage which
is a miniature water-cooled furnace.[24] Such a furnace has been used to
observe heated samples of pyrotechnic systems.[25] The furnace is heated
by means of nichrome wire, and temperatures are determined with a
plate-type thermocouple (platinum vs. 13% rhodium–platinum alloy)
connected to a potentiometric recorder. The sample is placed in a 6-mm
flat-bottom Inconel or platinum pan such as is used in differential thermal analysis (DTA). The pan is placed on the thermocouple plate in the
furnace, which is provided with a quartz window in the lid. The lid can
be rotated through 360° while maintaining a gas-tight seal so that condensed reaction products can be removed from view.[25]

The sample is viewed by reflected light through a stereomicroscope
fitted with a trinocular head for photomicrography. The illumination is
from a 150-W quartz–iodine light through a fiber optics cable at an
effective angle. An event marker is used to record every exposure on the
temperature record. Agfachrome® 50L professional color reversal film
(50AS) is usually employed.[25]

11.5.   COLD STAGES

Some up-to-date heatable stages are also coolable by means of a
cold gas[18,19] or liquid[18] circulating through the stage. Homemade cold

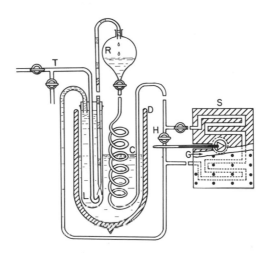

FIGURE 11.19.   Diagram of cold stage and cooling system.[26]

stages supplied with circulating liquid have been described[7,26] and employed.[27] In the early days, the big problem in circulating a liquid such as alcohol was in the pumping system. Figure 11.19 is a diagram of a system employing an airlift and a shunting arrangement of stopcocks so that circulation of liquid can be maintained after the stage is cold enough. Today there are commercially available miniature electrified mechanical pumps[28] which will not freeze up with such cold liquids. They are welcome successors to the bulky inefficient airlift.[26]

A newer development may be useful in cooling the liquid by means of a midget mechanical refrigeration system[29] (instead of dropping solid CO_2 into the liquid coolant[26]). The miniature refrigerating unit is capable of producing temperatures down to $-25°C$ in volumes up to 8 liters.[29]

Another method of cooling the liquid for a cold stage may be the thermoelectric cold plate, which cools $40°C$ below the temperature of tap water. For example, using $+20°C$ tap water cools the plate to $-20°C$; with $+10°C$ tap water it cools to $-30°C$; with refrigerated heat exchange fluid the plate cools to $-40°C$.[30] Other companies have combined a small thermoelectric unit directly with a cold stage, as shown in Figure 11.20.[31,32] Some other cold stages are cooled with a cold gas.[32]

11.6.   OTHER SPECIAL CELLS AND CUVETTES

There are many qualitative and quantitative kinds of special microscopical cells. Among the *quantitative* kinds are those of *specific volume*

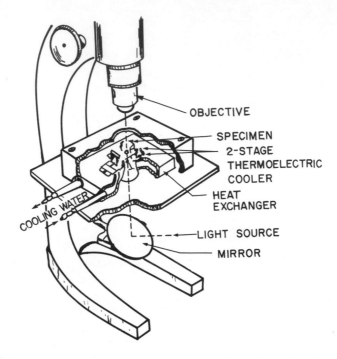

FIGURE 11.20. Microscopical cold stage with built-in thermoelectric cooler.[32]

such as *hemacytometer* cells, 0.10 mm deep with coordinate squares 0.0025 mm² engraved on the bottom.[7]

The Nichols stage refractometer is a set of two cylindrical cells cemented to a metal slide.[3,31] Each cell is equipped with two glass prisms (Figure 11.21). One set of prisms has a refractive index of 1.52, and is for liquids having refractive indices between 1.40 and 1.65. The

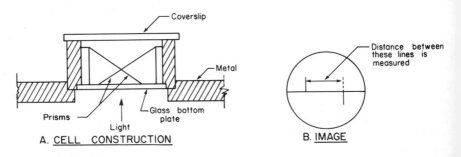

FIGURE 11.21. Nichols stage refractometer.[3] Courtesy of Microscope Publications, Ltd.

second set has prisms of 1.72, and is for liquids with refractive index 1.85 and higher. In calibrating a cell it is filled with liquid of known index, and the distance between two lines as seen through the microscope is measured with an eyepiece micrometer. The ocular distances for other standards are all plotted against the refractive index at a constant temperature to obtain the calibration curve at that temperature.

Colorimetry may be performed to some extent under the microscope by means of special cells and a photoelectrometer.[7] Polarimetry on liquids can be performed in a polarizing microscope, provided that it has a clear straight barrel that will take a long-enough tube filled to a known length to contain the liquid specimen. In this case a cap analyzer, rotatable in a ring, is used to measure the degrees of rotation of polarized light to the left or right by the fluid in the microscopical tube.[7]

There are also many kinds of *qualitative* cells, most of them auxiliary to the usual stage of the microscope. Illumination by radiation in the near UV requires UV-transmitting slides and perhaps cover glasses. Far ultraviolet or infrared radiation requires quartz or fused silica slides and cover glasses.[7] Qualitative microscopical analysis for any elements that are in both the specimen and in the ordinary glass slide requires the use of a special slide, such as a plastic one.[33]

Orthogonal unilateral illumination, as in slit ultramicroscopy, requires a special cell for holding the fluid which is to be illuminated from the side.[7] A similar cell fitted with electrodes may be used in *microelectrophoresis.*[7]

Pressure cells are required in the microscopical study of some technological problems, such as those dealing with aerosol cans or with polymerization under pressure. The principal problem is in specifying and testing the glass windows.[34]

In *biology* and allied technology there are many types of microscopical cells, chambers, and cuvettes for specimens *in vivo* or *in vitro*— stationary, moving, or changing with time. Some fundamental designs are shown in Figure 11.22.[35] Some of these cells are to be used vertically, and therefore the microscope is used horizontally.[34]

11.7. SUMMARY

Every microscope has a stage on which to examine the specimen. As in a theater, the stage may be stationary and rectangular in shape or it may be circular and rotatable. At the present time, polarizing microscopes usually are equipped with circular stages for rotating the specimen between the polars. However, there is some advantage to the

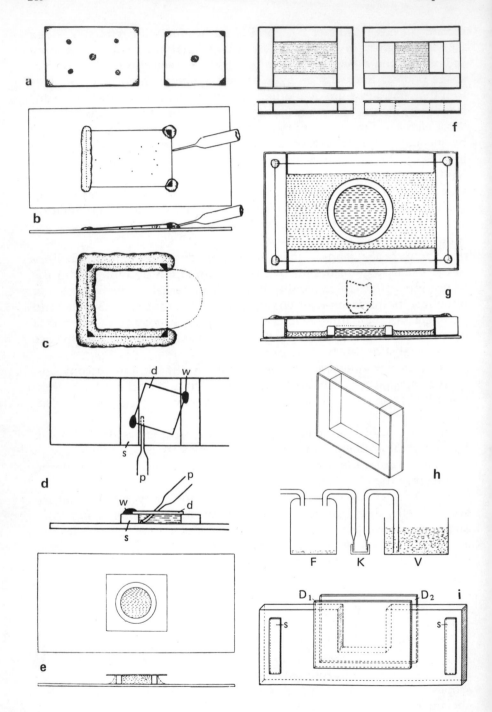

old-fashioned polarizing microscope with stationary rectangular stage and synchronously rotatable polars. The rectangular stage is relatively easy to provide with X and Y coordinate mechanical movements either at the factory or by means of an auxiliary attachment. The rectangular shape is also relatively easy to fit with other auxiliary stages such as for heating or cooling, micromanipulating, testing, or otherwise treating the specimen.

Heatable stages provide for the study of physical–chemical changes such as melting, crystallization, and polymorphic transformations. Commercial heatable stages are appreciably thick, presumably to provide adequate insulation, heat distribution, and chamber space, and therefore the microscopical objective must have an appropriately long working distance. Hot stages on the market which are limited to use with ordinary objectives of NA = 0.25 (10×) are adequate for most work requiring heat. However, to obtain interference figures, dark-field illumination, or certain other facilities, a higher NA is required. For these purposes special objectives may be used with both long working distance and adequate NA (0.85) for obtaining interference figures. Such objectives are provided by the manufacturers of universal stages, requiring the placement of a glass hemisphere about 1 cm in radius above the specimen.

FIGURE 11.22. Typical cuvettes for microscopical specimens *in vivo* or *in vitro*.[35] (a) Support for cover glass provided by droplets of paraffin wax (after Kuhl, 1949). (b) Wedge-shaped cell viewed from above and from the side, designed to prevent damage to very tender specimens by weight of the cover glass. At the left a wall of paraffin holds the edge of the cover glass; at the right paraffin legs and pads support and hold the opposite corners of the cover glass (after Kuhl, 1949). (c) Three-sided enclosure, of special advantage for living material. The reserve supply of water (drop extending at right) can be replaced as needed, to prevent drying of the sample and unnecessary motion of the liquid (after Kuhl, 1949). (d) Staining bridge, shown in top view and side view. The glass blocks are cemented to the slide, the cover glass d is fastened to them by clumps of wax w, and p is a pipette used to transfer the staining and washing liquids (after von Wasielewski and Kühn, 1914). (e) Ring-shaped cell for larger living specimens, shown in top and side views. A mantle of water surrounds the ring, preventing evaporation of the experimental liquid (after Kuhl, 1949). (f) Adjustable cell for larger specimens made of glass blocks cemented to and covered by photographic film, shown in top and side view (after Kuhl, 1949). (g) Combination of moist chamber and ring cell for larger, living specimens or monitored sampling, made of base plate, cemented glass blocks, and large cover glass fastened at the corners with paraffin (after Kuhl, 1949). (h) Above: Simple cuvette. Below: Arrangement for constant renewal of culture medium in the cuvette. By siphon action the liquid streams from the supply vessel V into the cuvette K, and is then sucked over into the flask F (after Bode, 1954). (i) Microaquarium after Schaudinn: The cutout in the glass block is enclosed on both sides by cover glasses D_1 and D_2. The blocks s are for protection (after Belar, 1928).

An alternative to the thick hot stage with high NA objective of long working distance is the hot stage which is thin enough to be used with ordinary objectives of NA up to 0.85. A thin hot stage may have to be especially built.[11] With a high-powered objective the thin stage is not recommended for temperatures much above 150°C without further insulation and consequent compromise in purpose.

Another alternative to the thick hot stage is the hot-wire stage.[23] This is especially designed to study phase transformations, including mesophases.[11] The wire may be heated to 300°C without apparent damage to the polarizing microscope. Temperatures are measured directly with a tiny thermocouple[5] or indirectly by observing the behavior of an adjacent standard system of phases.[23]

Very hot stages for temperatures of 1000°C and above are really miniature water-cooled furnaces. They require the long working distance of an ordinary stereoscopic microscope, a simple microscope, or a telescope. Some conventional hot stages for use below 300°C may be cooled by a liquid[18] or gas.[19] Cold stages specifically designed for cooling below room temperatures are commercially available. At least one[32] has a built-in thermoelectric cooler. Some others[29–31] have external thermoelectric devices and means of measuring temperatures. Of course, cooling a liquid with solid CO_2 and pumping this through the stage remains a useful method.[26]

There are many kinds of special slides, chambers, and cells used in microscopy. Plastic slides are substituted when ordinary inorganic glass or special silica slides are inappropriate. Hollow slides are used to hold more liquid than will stay in place by viscosity or capillary action. For quantitative determination of particles in a known volume of liquid such as blood, hollow slides are manufactured with precise cavities, sometimes with micrometric engravings on the bottom. The Nichols stage refractometer is a cylindric cell containing two crisscross glass wedges of known refractive index. When the cell is filled with the liquid of unknown refractive index, the measured distance between two bright refraction bands indicates the refractive index of the liquid.

In biology there are many other types of microscopical cells, chambers, and cuvettes for specimens *in vivo* or *in vitro*.[36] The specimens may be stationary, moving, or changing with time. Some of these cells are filled with liquid that will run out unless the cell is kept vertical and the microscope is used horizontally.

12

Transmission Electron Microscopy

12.1. ELECTRON MICROSCOPES

An electron microscope is an optical device for producing exceptionally high resolution of detail in the object by a beam of electrons.* There are three principal kinds of electron microscopes, classified according to how detail in the specimen is revealed by electrons: transmission, scanning, and emission. In the first two types free electrons are discharged from an electron gun to act upon the atomic nuclei of the specimen, whereas in the field-emission type the specimen itself is the source of radiation.[1] The field-emission microscope, employing no lenses, will be discussed in Chapter 14. The transmission electron microscope and the scanning electron microscope (see Chapter 13) employ focusable lenses.

The focusability and extremely short wavelength of electron beams are responsible for the theoretically high resolving power of the TEM and SEM. The rapid technological development of practical resolution, contrast, and other attributes contributing to visibility account for the great utility of these electron microscopes. Whereas the numerical aperture of electronic lenses is still relatively low, a compensating factor is their great depth of focus. Electrons react with the various atomic nuclei in the object rather than with the much larger domains in light microscopical specimens, so that electron microscopical images give unique information. Moreover, the preparation of electron-microscopical specimens is often different from the preparation of light-microscopical specimens. Different preparations mean different changes in a

---

* An electron is a subatomic particle having a negative charge of 4.8025×10^{-10} esu and a charge-to-mass ratio or specific charge of $5.2737 \pm 0.0015 = 10^{17}$ esu/g.[1]

specimen, which is to be taken into account in interpreting differing images.

In the *transmission* electron microscope (TEM) the image is formed from electrons which pass *through* the specimen. The resultant beam will contain some of the original free electrons which have not been changed in velocity or direction and some electrons which have been changed either way, or both.[2]

The transmission electron microscope is both a projection microscope and a photomicroscope. Since the electron image cannot be viewed directly by the eye, the image is projected onto a fluorescent screen and, when ready, is transmitted to a photographic plate or film. As such, the electron microscope is to be compared optically with the light-projection microscope or photomicrographic camera, as shown in Figure 12.1.[3]

A beam of free electrons emitted in a vacuum by a pointed filament F in Figure 12.1 can be condensed by an electromagnetic (or electro-

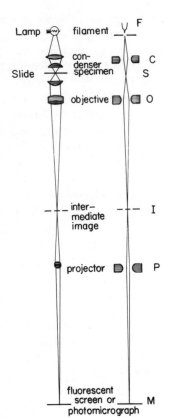

FIGURE 12.1. (left) Light-projecting microscope and (right) electron microscope, shown schematically.[3]

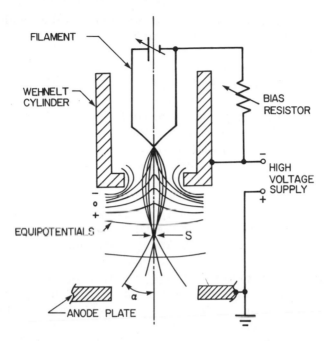

FIGURE 12.2.   Self-biased electron gun, showing lens-type constriction of the beam to a focal spot S ($\alpha = \frac{1}{2}$ angular aperture) [from *Practical Scanning Electron Microscopy*, J. I. Goldstein and H. Yakowitz (eds.), Plenum Press, New York, 1975, adapted from Hall[2]]

static[4]) lens C to an even smaller spot (2 or 3 μm) on the specimen S. Another electron lens, the objective O, focuses the transmitted beam to an intermediate image I, which is enlarged by the projector lens(es) P to form the image on the fluorescent screen or photosensitive material M.

The beam of electrons is generated and accelerated in the typical electron gun[1] shown in Figure 12.2.[2] The Wehnelt cylinder of the gun acts as a cathode shield and the base has a circular aperture (1–3 mm in diameter), centered with the filament's apex. Most modern electron-microscopical guns have a bias resistor applied between cathode shield and filament. They are actually self-biased because the negative potential between shield and filament is produced by the flow of beam current through the bias resistor (rather than by a battery). As with an unbiased gun, the distance between the tip of the filament and the aperture of the gun is adjusted to yield the most intense spot as seen on an intermediate screen or on the final screen at low magnification. With the biased gun the adjustment is much easier. Moreover, the filament's tip can be focused as a tiny single spot, resulting in a much higher intensity for a given beam current. Another distinct advantage is the relative insensitivity to fluctuations in beam current.[2]

12.2. ELECTRON LENSES

The chief lens in electron microscopes is the *objective* because it determines the degree of resolution in the image.[2] A simplified scheme of a magnetic objective lens is illustrated in Figure 12.3.[5] Whereas the iron yoke Y and electrical windings W are relatively large and externally impressive, the important pole pieces P_t and P_b have very precise bores B, an aperture A, and specimen S. The specimen may be located between the two pole pieces, or it may be slightly above the top pole piece P_t. As long as the distance above P_t is greater than a certain minimum, the distance is not precise and so the specimen may be tilted through large angles. Whereas there is less angular tolerance if the specimen is placed between pole pieces, there are the advantages of great tolerance of lateral motion, a simple mechanism for inserting the specimen, and shorter attainable focal lengths. But longer focal lengths give better contrast. An aperture A is usually inserted to increase contrast by cutting out scattered electrons. Of course, the smaller the aperture, the lower the resolving power and the more difficulty in keeping the microscope clean. Stigmators S_t to correct for spherical aberration are devices (usu-

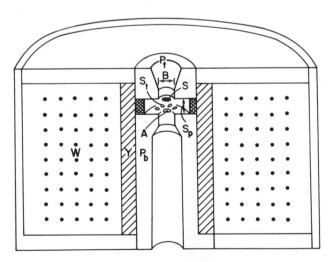

FIGURE 12.3. Simplified scheme of an electromagnetic objective lens with iron yoke Y, electrical windings W, pole pieces P_t and P_b, aperture A, specimen S, space between pole pieces S_p, stigmators S_t.[5]

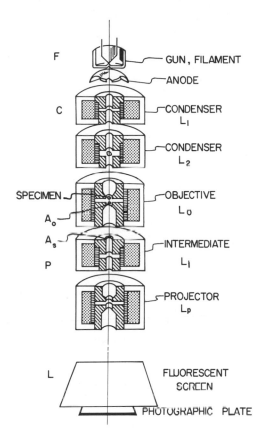

F — GUN , FILAMENT

ANODE

C — CONDENSER L_1

CONDENSER L_2

SPECIMEN — OBJECTIVE L_0

A_o

A_s — INTERMEDIATE

P — L_1

PROJECTOR L_p

L — FLUORESCENT SCREEN

PHOTOGRAPHIC PLATE

FIGURE 12.4. Scheme of transmission electron microscope (TEM) with double condenser and double projector.[5]

ally 6 or 8) placed around the space Sp and between the two pole pieces P_t and P_b to adjust the electron rays along the respective azimuths.

The *projector lens* is required to receive the small stream of electron rays and produce a relatively large image. In order to gain more variation in magnification, the projector lens system is usually a doublet[2] (see Figure 12.4[5]).

The purpose of the *condenser* is to form an image of the point of crossover between the annular rays from the aperture of the electron gun. Again, a double lens adds flexibility to the illuminating system in terms of intensity vs. area of the illuminated specimen[2] (Figure 12.4[5]).

In the 1940s there was sometimes a naive acceptance of the shiny new electron microscope as an instrument of miracles. The dangerous confusion of images with artifacts was overlooked in a blind enthusiasm

FIGURE 12.5. Electron microscope: worker of wonders. Drawn by C. C. Cory.

for this worker of wonders equipped with its own eyes and brain (Figure 12.5).

By the 1970s an awareness of possible confusion in image interpretation had become more and more apparent. Witness the ironic "ultrastructural sonnet" published in 1975 in the *Journal of Microscopy*[6]:

> In sombre beauty in her room she broods;
> 'Tis night—and all her pumps are deathly still,
> And thus she slumbers peacefully, until
> The morn, when unkind amperes end this interlude.
>
> With steady beat her motors wheeze and keen,
> Industrious vapours drain the inner core
> That Bohr's electrons shortly will explore,
> In headlong torrent downward to her screen.
>
> What truths does she uncover with her beam?
> How much is artefact produced by man,
> And how much really fits into the plan
> Of nature? That believed is easily seen!
>
> But even if she may promote confusion,
> It is at least an elegant illusion.

12.3. RESOLVING POWER

For *resolving power* the reasoning is the same as with light microscopes. Resolving power is fundamentally what you *pay* for: the poten-

TABLE 12.1[7]

Beam voltage (kV)	Wavelength, λ (Å)	Resolving power, d (Å)
50	0.054	2.1
100	0.037	1.7
1000	0.009	0.7

tial ability to distinguish between two points, expressed as a minimum distance. Hall pointed out[2] that this reasoning is not strictly logical, since "we say that the resolving power is higher the smaller the distance." He preferred the definition used in spectroscopy: Resolving power is inversely proportional to the minimum wavelength difference between spectrum lines.

For the *approximate* minimum resolving power as distance, d, we can borrow the theory from light microscopy: $d = 0.6\lambda/\sin\alpha$, where λ represents the wavelength and α equals one-half the angular aperture.[1]

Ruska[7] has estimated "attainable" resolving powers for some beam voltages (kV) and consequent wavelengths (λ). For the common voltages 50 and 100 kV and the especially high voltage, 1000 kV, his values are given in Table 12.1.

High-voltage TEMs such as the one shown in Figure 12.6 are built to attain the limiting resolving power and to test their practicability[7,8] (see Figure 12.7[9]).

FIGURE 12.6. 750-kV electron microscope of the Electron Microscope Section in the Cavendish Laboratory, Cambridge University, Cambridge, England.

FIGURE 12.7. Commercial (drawn) polyethylene terephthalate filaments etched with 42% aqueous solution of n-propylamine at 30°C, showing stress corrosion and extremely fine etching. Taken at 1000 kV on a high-voltage electron microscope.[9] The filament in the lower left is quite in focus; the central filament manifests the halo of underfocus.

Abbe had postulated that in the light microscope only one of the two diffracted rays of the first order is necessary (in combination with the ray of zero order) to form an image of a *periodic* structure. Under this condition, a diffraction grating could be twice as fine as the limit resolved by axial rays. The required double angle of aperture, 2α, can be achieved by tilting the illuminating beam. Gold crystals have a periodic structure of atoms separated by 2.06 Å (or 0.21 nm), and they have been resolved by tilting the beam to 2α.[2]

12.4 RESOLUTION

Resolution is what you *get* with your microscope from day to day. Practical resolution is dependent upon the value and constancy of the voltage as it controls the monochromatic wavelength, λ, and is inversely proportional to the angular aperture, α. The objective's aperture depends not only on the size of the diaphragm as delivered, but on the size before or after cleaning, as the case may be. Resolution is reduced from 0.3 nm when the objective has a focal length of 2 mm to 3.5 nm with a focal length of 14 mm.

12.5. CONTRAST

Contrast in the electron-microscopical image is a function of the nature of the specimen and the properties and adjustment of the electron optical system. Hall[2] takes up six factors: contour fringes, amorphous scattering, diffraction contrast, refraction, deflecting fields, and instrumental factors.

Contour fringes are most prominent at edges bounded by free space. They also occur at edges of dense particles embedded completely in a matrix of lower scattering power. At true focus contour fringes are invisible, but deviation from true focus affects resolution adversely.

Amorphous scattering is a function of sperical aberration resulting in the loss of part of the incident intensity from the imaging beam. Such scattering increases contrast as if a smaller diaphragm were used on the objective. For the same reason, resolution is lost as contrast is increased by amorphous scattering in spherical aberration.[2]

Diffraction contrast occurs in crystals as interaction between diffracted and undiffracted waves, as well as multiple scattered beams. It is a complicated phenomenon that is manifested as fringes at edges, extinction contours at grain boundaries, and curious patterns of dislocations.[2]

Refraction contrast occurs at boundaries between refractive material and a vacuum as a result of interference (phase delay) between the respective waves. The resulting phase contrast is manifested by over- or underfocusing.[2]

Deflection of beam electrons by a magnetic or electrostatic field is an especially important source of contrast. Boundaries between ferromagnetic domains are manifested in this manner. The phenomena in electrostatic fields in highly resistant materials are more complicated and may be erratic.[2]

The chief instrumental factor affecting contrast is the beam voltage. Decreasing the voltage increases the contrast, but the thickness of the specimen should be decreased accordingly; thus there is a very practical lower limit to the voltage. Contrast may also be increased by reducing the aperture of the objective. The limit to this procedure is set by the degree of diffraction error to be tolerated. Another limitation is that the smaller the aperture the sooner it will be plugged by contaminant[2] (chiefly, a carbonaceous product of charring organic matter). (See Section 12.7.)

12.6. ABERRATIONS

Aberrations in electron-microscopical images may be caused by either mechanical or optical problems. Mechanical problems arise from difficulties in machining pole pieces precisely and from inhomogeneities in the iron pieces. Either or both difficulties are manifested in astigmatism, i.e., appreciable lack of symmetry about the optic axis. In modern instruments, correction is effected by stigmators whereby the azimuth and magnitude of correcting asymmetry can be applied while the microscope is operating. A mechanical method uses two compensating iron pieces. Their distance from and along the optic axis may be varied until correction is accomplished. Another method uses electrostatic electrodes equally arranged around the optic axis. The azimuth and direction of the correcting asymmetrical field are controlled by separate potentiometers. The electrostatic method has no moving parts or vacuum seals, and the appropriate compensating circuit can be switched-in simultaneously with change in beam potential. Stigmators are used on objective and condenser lenses.[2]

Aberrations resulting from the projector lens system do not affect resolving power but may produce distortion in the final image, especially near the outer zones of the lens. The obvious "correction" is to avoid going to the limit of magnification that is set by the bore of a particular pole piece. The temptation to overextend the limitations of the projection lens is not as great in the modern two-lens projector system as in the single lens. It permits a greater range of magnification without exchange of pole pieces.[2]

Hall lists seven *optical* aberrations from electron (as well as light) lenses.[2] Among them all, axial–spherical aberration is the most important. This is the only theoretical aberration which occurs for a point on the optical axis as well as for all other (extra-axial) points in the specimen. The explanation is that from *any* point in the object the power of the lens is greater the longer the distance at which rays pass through the lens. Electron lenses are *always convergent;* the power of outer zones of the lens is always too great compared with that of the inner zones. Since, at present, there are no divergent electron lenses to combine with convergent ones, correction for axial–spherical aberration cannot be accomplished with electrons as it is with light. The tendency, then, is to keep the angular aperture small in order to reduce the aberration, and this limits the resolving power.[2] Ruska,[7] and Septier[10] are attempting to overcome spherical aberration and to attain the theoretical limit of the transmission electron microscope.

Some varieties of spherical aberration peculiar to electron *projector lenses can be corrected.* They are manifested as either pincushion or barrel distortion. In either case, the image lies on a curved surface so that the distortion occurs when projected on a flat surface. Correction is by means of a second lens which tends to neutralize the spherical aberration of the first. The pair can be either a doublet or a two-lens system.[2]

12.7. CLEANLINESS

Cleanliness in electron microscopes involves taking precautions against contamination and removing whatever contamination has nevertheless accumulated. A contaminant can be any product of bombardment by the intense beam of electrons in a high vacuum. The chief kind of contaminant is any carbonaceous product of decomposed organic matter: commonly grease used as vacuum seals, or oil leaking in as vapor from the forepump. Trouble from grease is avoided by using greaseless seals such as dry, nonvolatile, elastomeric gaskets, metallic bellows, a grease-free specimen, and photographical locks. Trouble from oil is reduced by having a remote forepump with a main valve to be kept closed until the diffusion pump is warmed up.[2]

Hydrocarbon vapors from oil, etc., decompose under electronic bombardment and are deposited as such or as polymers. Silicone oils may also deposit silica. If either kind of oil vapor is trapped away from the specimen, water vapor may become the chief offender by becoming ionized and etching any organic specimen. The apparent source of water is photographic negative material,[2] which therefore should be thoroughly desiccated before use.

Of prime importance is the protection of the specimen itself from contamination. Precautions should also be taken scrupulously to clean containers, instruments, reagents, substrata, replicating materials, and room atmosphere. Reagent and wash water should be freshly distilled, not stored in a metallic tank and run through pipes.

Cleanliness also means having clean lenses, open apertures, etc., just as in light microscopy.

12.8. DEPTH OF FOCUS

Depth of focus means, in practice, how far the viewing screen or photographic plane can be moved along the axis of the microscope and still receive satisfactory focus in the image. This distance is relatively long. Hall estimates that the viewing screen and photographic plate or film could be separated successfully by several inches.[2]

12.9. FOCUS

Focus is the point at which rays originating from a point in the object converge or from which they diverge under the influence of a lens.[1] As in light microscopy, an outline of a particle is imaged most sharply when focus is exact, though contrast may then be poorest. Sometimes an over- or underfocus halo may be preferred or inadvertently obtained (Figure 12.7).

12.10. ILLUMINATION

Illumination in the TEM is by transmission of electrons through the specimen. The source of electrons usually is an incandescent filament in the shape of a hairpin. Whereas its relatively large size gives mechanical stability and long life, its hairpin shape is surrounded by an asymmetrical electrostatic and electromagnetic field. A pointed filament is much better electron-optically but it is, of course, more fragile and has a shorter life.[2]

The condenser in the transmitting electron microscope serves principally to control the size and therefore the intensity of the illuminated field of the specimen. Even higher intensities may be obtained electronically by means of a *double* condenser. The second condenser also allows for greater reduction in the area illuminated, thus protecting the specimen outside the field of view from injury by the illuminating beam.[2]

Intensity of illumination is expressed in amperes per square centimeter, i.e., in the illuminated area of the specimen. Intensity is inversely proportional to the square of the magnification. Brightness may be defined as the flux per unit area per unit solid angle, usually measured in amperes per square centimeter per steradian.[2]

Tilted illumination, as mentioned before, is a way to increase resolving power. With the TEM it is also a method of obtaining oblique illumination. Another way is to put an eccentric diaphragm before the specimen. Either way, a bright image is sought against a dark background, especially with crystalline substances which manifest diffraction images.[2]

Radiation illuminating the electron-microscopical specimen is of constant wavelength to the degree that the accelerating potential on the tungsten filament is kept constant.

12.11. ANISOTROPY

Anisotropy can be considered in electron microscopy because the motion of electrons is somewhat analogous to the behavior of light in anisotropic materials.[2] Moiré patterns may furnish visibility of the structure of crystalline and other anisotropic specimens. For example, Holland[11] has demonstrated symmetry in a crystal of polyethylene lying on a substrate of crystalline polyethylene. The Moiré pattern is the same in two opposite quadrants and different from the pattern in the other pair of quadrants.

12.12. USEFUL MAGNIFICATION

Useful magnification in electron (as well as in light) microscopy is the minimum magnification required to show the desired detail in a specimen. Of course, the finest detail may be beyond the practical limit of resolution by means of electrons. In such cases, it may be better to define empty magnification, beyond which no new information is revealed.[1]* The required resolution must be in the electron micrograph before enlargement, and the enlarger must be in sharp focus on the original micrograph. Moreover, photographical copying procedures generally lose resolution.

12.13. FIELD OF VIEW

Field, the portion of the object in view, is very small in transmission electron microscopy—on the order of a few square micrometers at 20,000× within the microscope.[3] At any given magnification, the field is at present limited by the low numerical aperture (≈ 0.002); greater apertures carry unacceptable aberrations and distortions. Other limiting factors are faintness of image for viewing or photographing, susceptibility of specimen to damage by more intense illumination, practical selection, and number of fields to be viewed.

Experience in the selection of field may come directly through repetitive work in a specialized field, such as virology. But in broad scientific, technological, or industrial areas, mapping out significant fields

---

* Some empty magnification, however, may be advantageous in making measurements.[3]

comes through long experience with macroscopy and light microscopy.[3]

Selection of relevant fields lies within the general science and art of sampling. The ability to discriminate a representative from a nonrepresentative field takes a great deal of experience, thought, and judgment.

12.14. ARTIFACTS

An *artifact* is a spurious image, one which does not correspond directly with the true microstructure of the original specimen.[1] Obviously, artifacts can result in serious misinterpretation of appearances. Avoiding or reducing artifacts depends on understanding how and why they become visible. Some electron microscopical artifacts represent damaging changes in the specimen during exposure to the beam of electrons. For example, films representing the specimen, substrate or replica may shrink, tear, or curl from absorption of energy from the electron beam. Textile or paper fibers, or biological tissues may swell or evolve gas. Microorganisms may change in size or shape.[3] Observation of variations in such changes with time of exposure or intensity of the electron beam should lead to the prevention of artifacts.

Other artifacts are introduced by contaminations such as those discussed in connection with the electron microscope itself. For instance, specimens of carbon black particles may be observed to increase,[3] resulting from the deposition of carbonaceous products of pump oil, gaskets, substrates, or specimens themselves.[2] Another result of contamination is drifting of the image. If, after cleaning the microscope and taking precautions against contamination from pump oil and gaskets, such artifacts recur, changes in the preparation of the specimen should be considered. If the specimen itself is volatile, encasing it in an evaporated film of metal or other stable element might help. At any rate, changes in the specimen caused by the high vacuum in the electron microscope should always be considered and constantly studied. Some specimens may require thorough preconditioning in a vacuum comparable to that of the electron microscope.

12.15. CUES TO DEPTH

Cues to depth of structure in an electron-microscopical specimen may be of several kinds. The built-in cue is the relatively great depth of field in reasonable focus, about 10 μm, which is about 10 times the depth of high-aperture light-microscopical objectives. The result is a more realistic three-dimensional appearance of the electron-microscopical image than in high-powered light microscopy.[3]

Shadows as a cue to variations in depth of specimen are very important, but are obtained in a much different way than in light microscopy because electron-microscopical lenses are not of high aperture. Consequently, oblique illumination cannot be obtained in the electron microscope by blocking off one side of the aperture of the condenser lens (as is done in light microscopy). Instead, the specimen is "shadowed" before it enters the electron microscope. That is, it is prepared by obliquely depositing a material, usually a heavy metal, at an angle. The metal is evaporated from an electrically heated filament in a vacuum at great enough distance from the specimen for the metallizing atoms to approach as a beam. A vertical projection of height h casts a shadow of length $l=h/\tan\theta$, where θ is the horizontal angle of obliquity. Thus the height h can be estimated by measuring the length of shadow cast at a known angle θ. In the positive electron micrograph such "shadows" appear light, making interpretations of depth difficult for the observer who is accustomed to viewing dark shadows. Therefore, a negative print is usually made from a transparent positive print[2] (Figure 12.8).

A third cue to depth is *stereoscopy*, made possible by the great depth of field in focus in the electron microscope. Two electron micrographs are taken with the specimen tilted at two different angles to the axis of the microscope. They are best viewed in a stereoscope. Most commercial electron microscopes are built so that the specimen may be tilted to the appropriate angles for stereoscopy.[2]

At the present time, the techniques of Fourier transforms have been applied with the TEM so as to image periodic structures of molecules in three dimensions with a resolution of 7 Å, nearly three times the best resolution previously attainable. Fourier microscopy combines data from low-contrast electron micrographs with data from electron diffraction. Various views of the specimen are obtained by having the specimen on a tiltable stage. From such projections an image of high resolution is constructed in three dimensions.[12]

12.16. THICKNESS OF THE SPECIMEN

Thickness of the specimen in transmission electron microscopy is very important. For optimum visibility the the specimen must be thin enough to allow electrons to penetrate some parts significantly and yet thick enough for other parts to absorb electrons preferentially. Even atoms of low atomic number scatter electrons appreciably. Therefore,

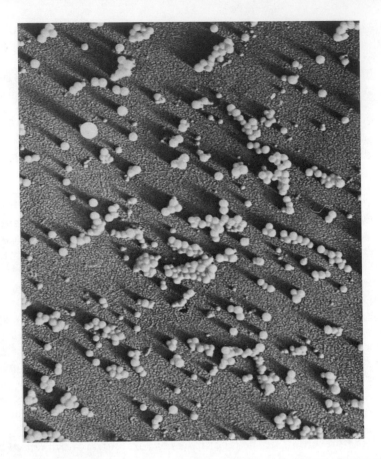

FIGURE 12.8. Polystyrene emulsion, stabilized and shadowed by coating lightly with gold (at an angle of 10°). A negative print used to make dark shadows. Electron micrography by Mrs. E. Gagnon Davis and E. J. Thomas, American Cyanamid Co.

sections of biological tissue and other organic materials must be much thinner than the thinnest specimens (1 μm) used in light microscopy. Inorganic materials have to be even thinner, if they are to transmit any electrons in contrast to a silhouette. The thinner or smaller a specimen, the more likely it is to be unable to support itself on the grid of the stage. Supporting films usually are no thicker than 200–500 Å (20–50 nm) in order to be sufficiently transparent to electrons.[13]

Particulate specimens such as dusts, precipitates, pigments, and strengthening "fillers" must be dispersed enough to differentiate units from aggregates. However, during preparation the size of the effective particle should not be affected if the electron micrograph is to be interpreted in terms of a problem such as lung disease, air or water pollution, pigmentation, or particulate strengthening.[3]

12.17. DEPTH OF FIELD

Depth of field in the TEM is sufficient to ensure that the whole field in view of an adequately thin specimen is in focus. To this extent, depth of field in electron microscopy is not a problem, but is instead an advantage in visibility. Of course, the depth of field varies with the angular aperture, 2α. While 2α is always small in electron microscopes as compared to light microscopes, the angular aperture and therefore the depth of field can be varied between limits. For a resolution d, of 10 Å, α is about 2.5×10^{-4} rad for beam potentials of 50–100 kV; the depth of field, D, is 4000 Å, and $D/2$ on the underfocused side is 2000 Å or 0.2 μm. For resolution of 50 Å, $D/2$ is 1 μm.[2]

12.18. STRUCTURE OF THE SPECIMEN

The *structure* of the specimen also influences visibility. For example, a *periodic* structure such as that of a single crystal of gold is resolved ($d = 2.04$ Å), when illuminated *across* the periodic structure by a tilted beam at an angle equal to the angle α of the objective. This resolution is five times the usual limit of 10 Å given above for a nonperiodic structure. The example is simply that of Abbe's diffraction principle, $d = \lambda/2 \sin \alpha$, provided that the undiffracted ray and one of the first-order diffracted rays are both included within the angular aperture of the objective. The resulting image is really a diffraction pattern of the (gold) crystal. It is a special example of how the structure of the specimen affects visibility. Nevertheless, it is a practical case, since all metals and their alloys represent periodic (crystalline) structures which can be resolved, provided they are illuminated by a properly tilted and directed beam of electrons.[2] The structures of other crystalline specimens also play an important role in visibility.[14]

12.19. MORPHOLOGY OF THE SPECIMEN

Morphology is the shape and size of particles, lines, areas, or volumes in a structure.[1] Seeing morphology is what microscopy is all about. Interpretation of shape is relatively easy when the size is large and the other attributes contributing to visibility are favorable. How-

ever, some microscopists seem to wish to see smaller and smaller particles or parts, regardless of larger neighbors having the same composition and shape. Other microscopists are not particularly interested in
cracking the barriers of resolution, but have no larger relevant particles
to look at. They would do well to seek or grow larger specimens. The
chemists's method of digesting the finest particles and precipitating the
resultant solute (in time) upon the larger particles might be useful. Or
sublimation or annealing might be tried. Otherwise, it may be worth-
wile to study analogous or isomorphous species of crystals. Likewise,
the biologist may learn relevant morphology by studying closely related
but larger species. The point is that morphology, true to its Greek deri-
vation, pertains more to shape than to size.

12.20. INFORMATION ABOUT THE SPECIMEN

Information is a very important attribute contributing to the interpretation of TEM images. If the required information does not come
with the sample, the electron microscopist should get what he feels he
needs. What is the problem? Is there macroscopical and light-
microscopical information relevant to the problem? How were the samples taken? Are they stable in a vacuum? Are they heat-sensitive? In
what are they soluble and insoluble? Are there good and bad examples
present or obtainable? Old and new? If there have been changes in
environment, treatment, etc., are there examples of each?

Likewise, the microscopist should furnish adequate information
about his electron-microscopical examination with every micrograph
submitted in his report. Magnifications must be accurately given and
usually are, but other essential information is sometimes omitted.
Which is the top side of the electron micrograph? Remember to orient
the picture so that the illumination appears to be from above, if you
wish the viewer to interpret shadows of elevations and depressions as
he would see them macroscopically. Also remember to tell him whether
he is looking at a positive or negative print, and how you made the
shadows dark. You may be educating your sponsor as well as informing
him, and in the long run the benefit rebounds to you. This means that
exact details of sample preparation should be given with advantages
and limitations of the method.

Procedures labeled "standard" should have been standardized by
an official society like the American Society for Testing and Materials.[15]
The standards of ASTM are the result of democratic deliberation and
voting of peers who represent the producer on one hand and the con-

sumer on the other. ASTM standards are published annually and recon-
sidered at least every 5 years. The Society's procedures are reproducible
by consultants and referees; the results constitute admissible evidence
in court. Bound notebooks of information about procedures, interpretations, illustrations, and experiments should be kept, as well as a filing
system of electron micrographs.

12.21. EXPERIMENTATION

Experimentation should be devised to test the interpretation of all
images that were not obtained by standard or repetitive procedures.
The microscopist's reputation for accurate interpretation is always at
stake. He and he alone knows how far his experience can take him, and
when he should take the time to vary his procedures in sampling, preparing specimens, and checking against contamination and aging of the
specimens. Other experiments involve gradient heating or cooling and
chemical or physical reactions. Admittedly, electron-microscopical procedures require a lot more time than with light microscopy. Second,
there may be a problem in convincing the supervisor that more experimentation is required immediately. Third, the customer may be the
most impatient of all parties involved.

12.22. PREPARATION OF THE SPECIMEN

Preparation of the specimen is too large a topic to be discussed in
procedural detail. The reader who wishes laboratory directions is referred to monographs on the subject[16-21] and to ASTM recommended
practices.[15,22] It must be emphasized here that any preparation of a
microscopical specimen changes it in some way and to some degree. In
transmission electron microscopy the primary purpose is to prepare a
tiny, thin specimen that will fit into the holder of the microscope. In
most instruments the holder is for a screen or grid with about 200 openings per inch. One popular shape of screen is a disk about 3 mm in
diameter.[2,13,16] In most cases the TEM specimen represents a very small
portion of the sample.

The first problem is whether to select a representative or nonrepresentative specimen. If we are interested in contamination, impurities,
forensic traces, spots, specks, or other isolated portions, *non*representative selection of the specimen is in order. In the process the original
sample has been changed drastically, so that a very accurate record of

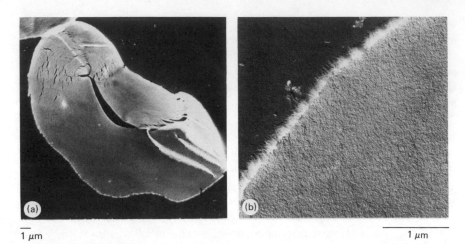

1 µm   1 µm

FIGURE 12.9. Transmission electron micrograph of a thin section of a cotton fiber.[23] (a) Cross section of whole dry fiber of cotton. (b) Higher resolution of area marked in (a). USDA micrograph by Southern Regional Center, New Orleans, La. 70179.

the selective process should be made. One good way is to photograph at a low resolution an area which will pinpoint the field taken at high resolution (Figures 12.9a and b[23]).

If sampling for TEM is to be *representative*, we have other considerations. Even a dry powder composed of only one phase but of small and large particles is difficult, if not impossible, to represent in a specimen of TEM size.[3] If the original sample is large enough, it may be advisable to make a quantitative size analysis macroscopically and then to take random specimens of the resultant size fractions for electron-microscopical illustrations. Size fractionation may be by rate of settling in a gas or liquid.[15,24] Settling in a liquid may be by gravity or centrifugation. Liquids and some gases can cause aggregation of particles. Unless such aggregates are penetrated by the electron beam, they will appear in silhouette and can be mistaken for single, much larger particles. Some liquids may partially dissolve particles; others may swell the particles. Without sufficient information it should not be assumed that dispersion in any one preparatory liquid will duplicate commercial dispersion in a different medium. Many a false interpretation of *practical* particle size or shape has been made by assuming the morphology to be that of the "ultimate"[2] (finest) size of the particles before introduction into the commercial product.

Often particles may be deposited directly onto a filmy substrate such as Formvar® polyvinyl formal resin by settling from a dispersion in air or by dusting or rolling from a cotton swab dipped into the powdery sample. Other powders may need more shear force, such as that provided by a Tesla coil operating on a small portion of the powder resting on a Formvar® film. An alternative is an ultrasonic vibrator operating on the powder suspended in acetone so as to spray the suspension onto a carbon film. Magnetic particles, or any other dry materials which would contaminate the TEM, must be embedded in an adhesive such as a 1% solution of collodion (cellulose nitrates) in amyl acetate. Dispersion is by means of a spatula or muller on a glass plate[16] or a tiny mortar and pestle. Enough solvent is added to leave a sufficiently thin film after evaporation. If aggregation occurs during this treatment, it may be better to use a vehicle and a thinner that represent natural or industrial conditions. Particles are usually "shadowed" by evaporating a heavy metal at a known angle[16] (Figure 12.8).

Surfaces are generally prepared for the TEM by replication. Carbon replicas are preferred because they are thermally and electrostatically conducting, and they can be made thinner than plastic replicas. A typical method of making a carbon-positive replica of a solid surface is shown in Figure 12.10. First a 4% solution of collodion in amyl acetate is applied, dried, stripped off, and discarded to clean the surface. Then several layers are applied and dried to build up a firm base. This negative replica is stripped off and shadowed by evaporation of a heavy metal on the replicating side. Next, carbon is applied as a positive replica from a carbon arc in a vacuum.[16]

Surfaces which are sensitive to the solvent of a plastic may be replicated dry by using polystyrene disks made by flattening pellets at 165°C. Powders, such as dry yeast, are especially amenable specimens; they are simply sprinkled on the disks, covered with a weighted glass slide, and placed in the oven at 165°C for about 5 min. After cooling, the specimen is removed and carbon is projected from a carbon arc in a vacuum, as usual. Since carbon replicas are very delicate, special techniques have been developed to handle them.[16]

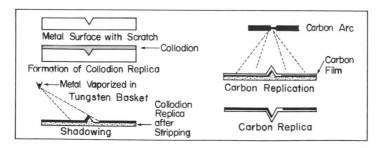

FIGURE 12.10. Steps in making a carbon-positive replica.[16]

Microtomy is a highly specialized technique for preparing thin sections of a specimen for the TEM (see Figure 12.9). Ultramicrotomes have been developed to advance the specimen very slowly, usually by thermal expansion of the metallic holder. Glass or diamond knives are used, and a relatively firm embedding medium is required. Epoxy resins are favored, but they require that the specimen and all that the resin touches be completely free of water. The ingredients must be consistently proportional and thoroughly mixed. The proportions are based on the producer's advice and the consumer's experience. The required 60°C heating for polymerization may be a drawback with heat-sensitive specimens. As a result, the butyl–methyl methacrylate monomer, polymerized by ultraviolet radiation at room temperature, has gained favor. The butyl methacrylate softens the methyl methacrylate in a ratio chosen from 9 : 1 to 4 : 1. With either embedding medium, biological and other hydrous specimens must be "fixed" and then dehydrated with ethyl alcohol, starting with a bath of 50% and going through 75%, 95%, and 100%.[16]

Freeze-fracturing biological and other aqueous specimens is accomplished by freezing the specimen in Freon® 12 refrigerant and producing a fracture surface with a precooled razor blade. The specimen is transferred to a special accessory for the Ladd vacuum evaporator, wherein at the precise moment the specimen is shadowed with platinum + carbon and replicated with carbon. The replica is protected with a drop of 1% collodion in amyl acetate, and a slice of the specimen plus replica is cut off with the razor blade. The specimen is dissolved away with Chlorox® sodium hypochlorite solution, and the collodion is dissolved away with fumes of amyl acetate, leaving the carbon replica for electron-microscopical examination.[16]

12.23. ELECTRON MICROGRAPHY

Accelerating voltages of 50–100 kV produce electrons each of which is capable of passing through as many as 100 photosensitive grains, losing enough energy to make at least some of the grains developable. The number of grains developed depends on the developer and the development time. The information in a ray of electrons is the "signal." There is also "noise" (without information). However, the *signal-to-noise ratio for electrons* is 100–1000 times greater than that for light! The photographic record also has a signal-to-noise ratio; the signal is the optical density, and the noise is the developed granularity.[26] If there is no granularity[1] developed, the signal-to-noise ratio in the electron micrograph is the same as in the electron beam. Such a film is on Kodak®

high-resolution plates[25] which are practically without graininess[1] but are far too slow for any except the most stable instruments and specimens.[26]

Among glass photographic *plates*, the "projector slide" grades are the most practical.[25] As far as electrons are concerned, there is little difference between the two grades, "medium" and "contrast." One great advantage to both grades is that useful exposure time to electrons (speed) can be changed by adjusting the conditions during development. Normally, Kodak® developer D-19® solution is recommended for 3 min at 68°F (20°C). However, development time can be varied from as little as 1 min to as much as 8 min.[26] Photographic plates have the greatest dimensional stability and are readily prepumped before use.[27] But the disadvantages of heavy, bulky, and breakable glass plates are obvious when it comes to storage, both before exposure and after processing.

Fortunately, Kodalith® LR film, Estar® polyester base as cut sheets overcomes the problems of glass plates. Moreover, the polyester base pumps down almost as fast as glass and much faster than acetate bases. Polyester also has less curl and more dimensional stability than acetate. Kodalith® LR film, when developed in D-19® for 2 min at 68°F (20°C), has the same electron-reaction speed as medium projector slide plates with 50-kV electron acceleration and two-thirds the electron-reaction speed with 100 kV. Because of gradual fogging, however, increased development is impractical, and a Kodak® safelight filter 1A (instead of 1) is required during all handling prior to fixation. Kodalith® LR Estar® film is available in all sheet sizes.

For TEMs equipped for 35- or 70-mm roll film, the fine-grain variety has been the most popular. A drawback for some people is that the acetate base requires four times as long to pump down as glass plates. The electron-reaction speed of this film is slower than that of projector slide plates, but this is acceptable because the "bellows" length is shorter and therefore the electron-beam intensity is higher. The signal-to-noise ratio is low enough to be just barely detectable by the eye; this characteristic may or may not be important. Kodalith® LR Estar® film is available in rolls 35 or 70 mm wide, but only on special order.[26]

Besides potential granularity in a photographic emulsion, there is the "spread function" which limits the acceptable extent of enlargement of TE micrographs. The spread function measures the area over which an electron's energy is expended in its deviated path through the gelatin (not the photosensitive grains) of the photographic emulsion. Two narrow electron beams separated by less than the radius of spread (scatter) will not record separate information. For accelerating voltages of 50–100 kV the spread function is about 5–10 μm, giving a maximum enlargement of 20×. Less than 20×, developed granularity limits the photographic resolution of information with all materials except Kodak® high-resolution plates, which have practically no granularity.[26]

After exposure to electrons and specific development of the negative, the stopping, fixing, washing, drying, and printing are similar to the processes used after exposure to light. The reader is referred to Chapter 8.

There is, however, a drawback in the photographical systems of most TEMs: they do not accommodate the fast self-developing, direct-positive camera backs such as Polaroid®. These cameras must be used outside the vacuum system of the TEM. Focusing on the outside of the fluorescent screen is unsatisfactory because of the granular nature of the screen. Instead, a grainless single crystal of europium-activated calcium fluoride is employed. The activated crystal, assembled with a mirror, is placed in one of the two portholes provided for accessories in most TEMs. As indicated in Figure 12.11, a special light microscope M, consisting of an objective camera lens ($f = 1.4$) and an ocular, is placed in the other porthole. Connected to the microscope M is a Polaroid® camera C, supported on a stand S. The fluorescent crystal's assembly F includes a vacuum-tight sleeve that allows for focusing on the fluorescent image by viewing with a hand magnifier held in the plane of the camera back. The sleeve F also allows the whole assembly to be drawn out of the way when not in use. Figure 12.12 is an electron micrograph of a diffraction grating taken as a test of resolution and calibration.[28]

With the TEM it is important to select an area to be micro-graphed before the specimen changes, burns up, or contaminates the microscope. This is especially difficult at first if the image is weak or low

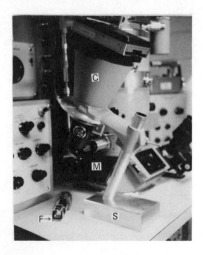

FIGURE 12.11. Polaroid® camera system attached to a TEM. A fluorescent crystal mirror assembly (F) is on table in foreground.[28]

FIGURE 12.12.   Micrograph of 1134 line/mm grating replica at 5000× magnification made with Polaroid® camera system of TEM.[28]

In contrast. While some TEMs are equipped with image intensifiers, others are not. For the unequipped microscope there is an accessory which intensifies both brightness and contrast of the image, and fits on the microscope in place of the binocular viewer. The accessory focuses the original image upon a photocathode and reproduces it on a phosphor associated with an anode. The net gain in luminance is about 4×10^4.[28]

The accessory image intensifier can be connected to a cathode-ray (video) tube. Such a system has been used to examine electron-microscopic particles from celestial space. Some of these particles decompose readily in the electron beam and therefore are examined at extremely low-beam intensities with the image intensifier.[29] The video aspect of the system is an apt introduction to the STEM.

12.24.   SCANNING TRANSMISSION ELECTRON MICROSCOPE

The *scanning transmission electron microscope* (STEM) combines the transmission of electrons through an appropriately thin specimen with the video-scanning technique which is characteristic of the scanning electron microscopy of surfaces. As shall be discussed in Chapter 13, the scanning system requires an especially powerful electron gun because each of the million scan points on the specimen is individually very weak.[30] An especially powerful source of electrons is the cold field-emission electron gun, shown in Figure 13.8. With the STEM the problems of instability of the gun and severe requirement of the vacuum

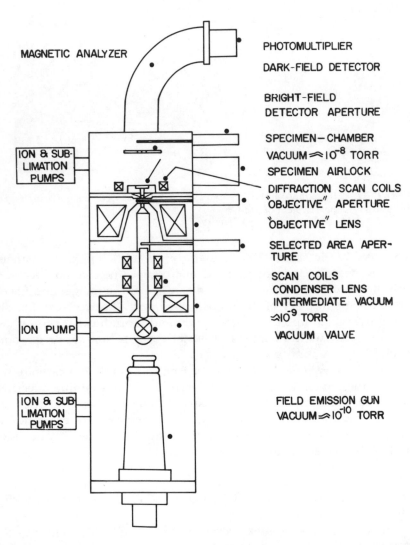

FIGURE 12.13. Diagrammatic sketch of scanning transmission electron microscope (STEM), inverted type.[30]

TABLE 12.2[30]

Fundamental advantages of field emission STEM relative to TEM	Advantages of convenience of field emission STEM relative to TEM
Simultaneous bright- and dark-field images	No change in focus with change of magnification
Elastically scattered electrons collected over a large solid angle	No image rotation with change of magnification
	Constant image brightness with change of magnification
Inelastic loss electrons separated by energy analysis with fixed object position	No upper limit to magnification
	Built-in diagnostic features
Electronic contrast expansion	Built-in ultimate performance image intensifier
Point-by-point signal processing	Convertible to high-resolution SEM
Elimination of chromatic aberration due to specimen energy absorption	Automated image analysis without loss of efficiency
Cold cathode source gives lower emission energy spread	Compatibility with scanning x-ray image analysis
	Long-life electron source
Irradiation limited to area of observation at all times	Simplified column alignment

have been solved.[30] Figure 12.13 is a diagram of the inverted STEM. The present authors have added quotation marks around "objective" lens and "objective" aperture because these items may be considered to be part of the condensing system. (They come before the object specimen rather than after.) (See Figure 12.1.)

The STEM microscope scans the thin specimen with a probe only 0.3 nm in diameter so that it compares favorably with the TEM in resolving power and quality of image. In addition, the STEM has certain advantages, as indicated in Table 12.2.[30] The dark-field image is produced by an annular collector of electrons, and the bright-field image is produced by a central collector of electrons. The two images may be presented simultaneously and can be analyzed in various ways. For example, if A represents the bright-field signal and B represents the dark-field signal, the ratios A/B, B/A, and A/AB can be computed.[30] Indeed, the STEM is sometimes designated as EMMA, electron-microscope microprobe analyzer.[31] Without any compromise to image formation, the STEM may be converted to provide energy dispersive x-ray microanalysis and electron energy loss spectrometry of thin sections.[30]

12.25. SUMMARY

Electron microscopy began with the realization that a beam of electrons behaves, like visible light, both as a train of waves and as a stream of particles. Since the de Broglie wavelength of electrons decreases with their kinetic energies, fast-moving electrons have very short wavelengths associated with them and so are capable of very high resolution if that wavelength can be brought to use in an appropriately designed instrument.

The transmission electron microscope (TEM) is one such instrument. It is arranged much like an ordinary microscope designed for the examination of translucent specimens by the transmission of visible light, except that magnets are used, instead of light-bending lenses, to deflect and focus the beam of electrons. The electrons originate in an "electron gun" that usually has a hot filament (sometimes a cold cathode emitter) as the source, plus an arrangement for defining and accelerating a narrow beam of electrons. The accelerating voltage may be anywhere from 30,000–1,000,000 V or more, but usually is in the range of 50,000–100,000 V. The accelerated beam is then focused on a tiny area of the specimen by a pair of concentric toroidal electromagnets which act as condensing coils. The sample is held on a screen of extremely fine wires, so the beam passes through one of the holes in the screen and on through the objective and projection electromagnets. Since the different parts of the specimen absorb electrons differentially, as the projected beam of electrons falls on a fluorescent screen it shows bright areas where the sample has absorbed least, and darker areas where the sample has absorbed more of the electrons. By removing the fluorescent screen and allowing the projected beam to fall on a photographic plate or film of the lantern-slide type, a latent image is obtained which can be developed in a photographic darkroom in the usual way.

Since the electron beam would be scattered by collision with air molecules, the interior of the electron microscope has to be pumped down to extremely low pressure by efficient diffusion pumps and forepump. It follows that any specimen which will dry out or lose volatile components in vacuum will be altered, and this limits the applicability of the method and tempers the interpretation of the image. Furthermore, volatile carbonaceous material will be decomposed by the hot cathode and by the stream of energetic electrons, depositing decomposition products in the fine bores of the coils and fouling the microscope. Decomposition of the sample by the energetic beam produces the same undesirable result. Volatile components of lubricants, sealing compounds, and pump oil are to be avoided for the same reason. Some contamination of the microscope is unavoidable, though, and periodic thorough cleaning of the critical parts is necessary.

The resolving power of a TEM is theoretically proportional to the wavelength of the electrons and inversely proportional to the angular aperture. The theoretical limit is never attained in practice, but a resolution of about 4 Å at 50 kV and 3 Å at 100 kV is attainable. Tilting the sample improves the resolution. In general, greater resolution is obtained on an electron photomicrograph than on the visual fluorescent screen in the microscope, because the photographic emulsion has much finer grains.

Contrast in the TEM image depends on many factors, but operationally the contrast may be increased by reducing the beam voltage. This allows greater differential absorption by the sample, but of course it reduces resolution. Similarly, reducing the aperture of the objective (by means of a diaphragm) increases contrast at the expense of resolution. Greater contrast in the photomicrograph can be obtained by using high-contrast plates and developer, of course.

Some aberrations of the magnetic lenses may be corrected by adjustable electrostatic deflectors within the TEM. Some progress is being made to reduce spherical aberration so as to approach theoretical resolution.

Depth of focus presents no problem in the TEM, being ample for all purposes. Deliberate under- or overfocus will increase contrast, but reduces sharpness.

The useful magnification of the TEM depends on its resolving power. At a resolution of 10 Å, a magnification of 50,000 is routine. The field of view under such conditions is very small, only a few square millimicrons. Hence the operator must usually choose between several kinds of field, based on experience, and must justify his choice. Recognizing spurious images as the result of damage to the specimen by the electron beam or fouling of the microscope is also a matter of experience.

Preparation of the specimen is an especially important part of operating the TEM. The sample must be thin enough to transmit electrons, yet thick enough to show differential absorption. Variations in depth can be enhanced greatly by "shadowing" the sample, depositing a heavy metal at an acute angle by evaporating the metal from a hot filament off to one side in a vacuum chamber. Further details of structure can be revealed by tilting the sample in two or more directions. Experiments can be conducted on the sample material to select a range of particle sizes, or to determine the effect of drying or heating, or to gain greater dispersion of solids, but of course any alteration of the sample material deemed necessary as a result of such experiments must be justified in the report. Many suggestions for sampling material, preparing a specimen, and interpreting the image are given, but for details see references 16–21.

The scanning transmission electron microscope (STEM) combines the transmission of electrons through the specimen, as discussed in this chapter, with the scanning and analytical systems of the scanning electron microscope (SEM), as shall be discussed in Chapter 13. The threshold current (flow of electrons) through each scanning point of the specimen is achieved as a useful signal by increasing the intensity of the source (gun). The power comes from the pointed tungsten tip, at room temperature, in a very strong electrical field and in a very high vacuum. The cold field-emission of electrons has been stabilized and maintained so that the resolving power of the STEM is comparable to that of the SEM. Moreover, bright-field and dark-field images can be obtained simultaneously. The corresponding signals can be put through a computer to obtain analytical data regarding very local losses in electron energy. Used this way, the STEM is also known as EMMA, electron-microscope microprobe analyzer. The STEM may also be modified to provide energy dispersive x-ray microanalysis.

Scanning Electron Microscopy

13.1. INTRODUCTION

In a scanning electron microscope (SEM) the image is formed in a cathode-ray tube synchronized with an electron probe as it scans the surface of an object.[1] The resulting signals are secondary electrons, backscattered electrons, characteristic x-rays, Auger electrons,* and photons of various energies (Figure 13.1).[2]

In a typical SEM an electron gun and multiple condenser lenses produce an electron beam whose rays are deflected at various angles off the optic axis by the first set of electromagnetic scan coils. The second set of scan coils deflects the beam back across the optic axis. Both sets of scan coils are in the bore of the final lens (not shown in Figure 13.1). All the rays pass through the final aperture (FA) of the final lens. From this final crossover, the rays, one at a time, strike the specimen at various points—for example, at positions 1 through 9 in Figure 13.1. The scan coils and the cathode-ray tubes (CRTs) are powered by the same scan generator, so that each scanned point on the specimen is unique as reproduced on the displaying or the recording CRT and video amplifiers. To these amplifiers are fed one or more of the resultant signals: high-energy backscattered electrons, low-energy secondary, and/or backscattered electrons, x-rays, and cathode-luminescent radiation in the ultraviolet, visible, and infrared regions. All the results can be monitored separately or simultaneously by means of the appropriate detectors.[2]

---

* Where electrons or x-rays bombard a specimen and ionize the atom, vacancies are formed in the inner electron shells. The atom rearranges by filling in with an electron from a shell of lower energy, and emits energy—some as x-ray photons and some as electrons of specific energy (Auger effect).[2]

The *interpretation* of scanning electron micrographs is different from that of images formed directly by bending light or electron rays from object to image. The SEM indirectly constructs a pattern or map that may be interpreted as an image of the object. Interpretation of the SEM "image" is promoted by the many other attributes contributing to visibility, particularly resolution, contrast, depth of focus, morphology (topography), apparent illumination, and three-dimensional aspect.

One of the most striking aspects of the SEM image by secondary and backscattered signals is the resemblance to images of depressions and elevations illuminated from above by an oblique beam of light (Figure 1.4). For example,[3] the cotton fabric shown by scanning electron microscopy in Figure 13.2 looks as though it were illuminated by an oblique beam of light from above. We have been familiar with this type of illumination in macroscopy since childhood. It is also the type of illumination which is used most frequently with simple magnifiers and stereoscopic compound microscopes. Consequently, the

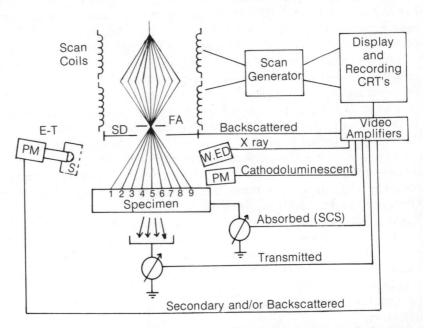

FIGURE 13.1. Image formation in the scanning electron microscope. FA, final aperture; SD, solid state electron detector; E-T, Everhart–Thornley electron detector; PM, photomultiplier; S, scintillator; W. ED, wavelength- and/or energy-dispersive x-ray detectors; CRT, cathode-ray tube[2]; SCS, specimen's current signal. Courtesy of J. I. Goldstein, H. Yakowitz, *et al.*

FIGURE 13.2.   Cotton fabric simply coated with a thin layer of gold + palladium by evaporation in a vacuum and photomicrographed by scanning electron microscopy.[3] USDA Photo by Southern Regional Research Center, New Orleans, La.

structure and morphology of the weave, the two yarns, and all the fibers are easily inferred.

Even when the magnification is boosted to 2000× so that the resolution and depth of focus are beyond that of the light microscope, a scanning electron micrograph is easy to interpret because it looks as though it were illuminated by a unidirectional beam of light from above the specimen[3] (Figure 13.3). Another aid to interpretation is the perspective these scanning electron micrographs manifest.

13.2.   RESOLVING POWER

The *resolving power* of the SEM depends primarily on the effective beam diameter of the probe. If it is assumed that the secondary collection efficiency is unity, and the specimen is capable of producing a contrast of the 25% necessary to satisfy Raleigh's criterion (Chapter 2), a theoretical limit of resolution of 90 Å (9 nm) is calculated for the ideal specimen.[2]

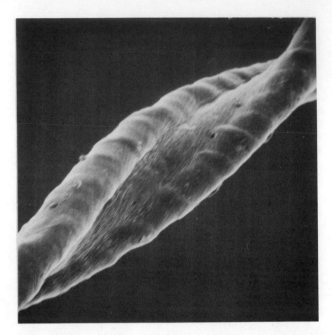

FIGURE 13.3. Cotton fiber, coated with a thin layer of gold + palladium by evaporation in a vacuum and photomicrographed by SEM.[3] Vividly shows thickening of flat fiber, fibrils, and two spiral structures. USDA micrograph by Southern Regional Center, New Orleans, La. 70179.

13.3. RESOLUTION

The practical *resolution* of the SEM is limited to 200 Å (20 nm) for most specimens in optimum position relative to the detector for secondary electrons. Some pairs of points on the surface of the specimen may not be in the optimum position for detection of secondary electrons, and therefore they are not resolved, even though the limiting pairs in optimum position are resolved[2] (Figure 13.3).[3]

Nonconducting specimens are coated with a metal to reduce damage by the thermal effects of the electron beam and to eliminate electric charges on the surface.[2] Thin as such coatings are (100–300 Å; 10–30 nm), they do cover up the finest topographical features, thereby reducing the practical resolution of the very finest detail.

13.4. CONTRAST

Contrast in scanning electron microscopy may be defined as the ratio of the change in signal between any two points on the specimen and the average signal. Differences in *atomic number*, i.e., nuclear charge, of the atoms composing the specimen, are important in the SEM, just as in the TEM. In the SEM, however, we are concerned chiefly with backscattered electrons rather than transmitted electrons (Figure 13.1). The higher the *atomic number* the greater the backscattering and the lesser the transmission of electrons (Chapter 12). For a given phase such as that of an alloy, ore, rock, ceramic, or cement, the effective backscatter coefficient is the weighted average of the backscatter coefficients of the pure elements. In smooth plane surfaces we are dealing entirely with atomic contrast[2] (Figure 13.4).

In the case of rough surfaces we are dealing with *topographic contrast*, whether backscattered or secondary electrons are detected. When a plane specimen is tilted away from normal incidence of the beam, the *backscattering* of electrons increases, gradually approaching unity at grazing incidence. Therefore a faceted surface (Figure 13.5) with its facets at various angles to the beam will emit signals of varying strengths. The peak in backscattering by each facet lies in a plane containing the normal to the surface and the direction of the incident

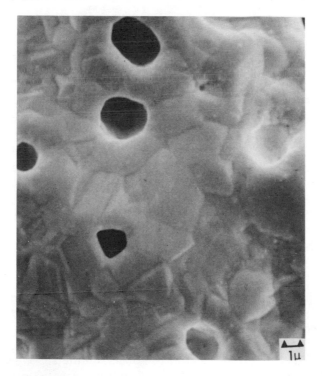

FIGURE 13.4. Pure gold wire after annealing to smooth surface (plus voids, black) at 950°C for 2 hr.[10] Courtesy of A. C. Reimschuessel.

10 μm

FIGURE 13.5.   Lead sulfide, galena (isometric), taken by F. G. Rowe with field-emission electron gun (Figure 13.8).

beam[2] (a sort of oriented "luster"). The Everhart–Thornley (E–T) detector (Figure 13.1) is highly directional and picks up only the trajectory electrons that are traveling toward the detector. Thus the resultant image contains a high proportion of black (shadowed) facets. If the specimen's current signal (SCS) (Figure 13.1) is employed to form the image, the trajectory effects are lost and the topographic contrast is due solely to the relative number of electrons from each facet.[2]

Secondary electrons can escape from the surface of most materials at depths of less than about 10 nm (100 Å). At such shallow depths there is minimal elastic scattering; the electrons travel in a beam parallel to the direction of incidence. A greater number of electrons will escape from surfaces at higher tilt.[2]

Contrast in the SEM can be varied in half a dozen other ways[2]:

1.   Electron channeling contrast, related to the crystallographic nature of such a specimen.

2.   Magnetic contrast related to crystalline structure.

3.   Magnetic contrast related to noncrystalline specimens.

4.   Voltage contrast, from beam-induced current contrast.

5.   Electron-beam-induced current contrast, a phenomenon of electron–hole pairs in semiconductor devices.

6. Cathode luminescences, from the emission of long-wavelength electromagnetic radiation of light. A very important advantage of cathodoluminescence is that it introduces natural *color* contrast into the SEM. Photomacrographs or photomicrographs may then be taken in color.[2]

Synthetic color may be introduced into a series of correlated black and white (e.g., x-ray) images of the very same area (e.g., an alloy) by arbitrarily assigning a different color to a different image.[2]

13.5. ABERRATIONS

Spherical aberration is caused in the SEM by electrons moving in trajectories which are away from the optical axis being focused more strongly than those near the axis (Figure 13.6, top). Consequently, the image of point P is a disk or "circle of confusion." The aberration can be reduced by diaphragming but with a lower intensity of the probe. Or the object can be placed nearer the lens, but with a loss in working distance and a loss of freedom from the condenser's magnetic influence.[4]

Chromatic aberration is caused by a variation in the energy from E_0 to $E_0 + \Delta E$ (Figure 13.6, center). Of course, a variation in the magnetic field will also produce a disk of confusion in the scanning probe, but sources of trouble can be reduced by stabilizing the respective power supplies. Yet there remains a fundamental variation, ΔE, in the energy representing the spread of energies leaving the cathode (2–3 eV for tungsten or lanthanum boride source; 0.2–0.5 for field-emission cathodes).[2,4]

Since there is wave motion in the electron beam and a diaphragm of radius R, *diffraction* causes its own disk of confusion, dr, in the scanning probe of electrons (Figure 13.6, bottom). The radius of the first diffraction minimum, dr, subtends an angle θ at the lens.[4]

Astigmatism enters the electron-probe lens by machining deviations, inhomogeneities within the iron pieces, asymmetry in the windings, and uneven accumulation of contaminants. Correction is usually supplied by a stigmator which has two controls: one for the *magnitude* of asymmetry and the other for the *direction* of asymmetry. Correction is by adjusting the two controls alternately and refocusing at medium magnification, 5,000–10,000×.[2]

13.6. CLEANLINESS

Cleanliness in the SEM is as important as it is in the TEM for the same reasons: to keep the apertures open and to avoid accumulation of

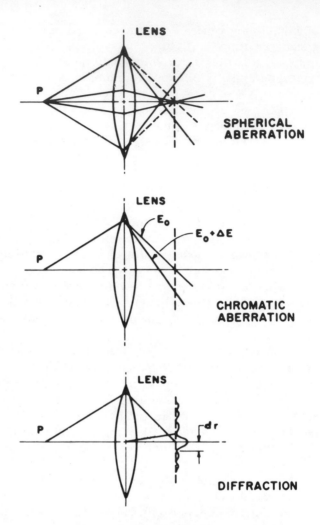

FIGURE 13.6. Schematic drawing showing spherical and chromatic aberrations as well as diffraction at a lens aperture[2] (adapted from Hall[4] and Oatley[5]).

carbon and of any other matter which may deviate the electron beams. Again the chief offender is carbon or carbonaceous matter from sealing grease, pump oil, or the specimen itself. Precautions, too, are the same: greaseless seals, cold oil-traps, and prompt removal of carbonaceous specimens. In Auger electron analysis,[5,6] an oil-free, ion pump is required.[2]

In x-ray analysis, contamination by hydrocarbons may lead to polymerization rather than decomposition. X-rays of long wavelength, such as from beryllium, boron, or carbon, may be highly absorbed, leading to a high degree of polymer-contamination and an increasing carbon $K\alpha$ count as a function of impingement time. One preventive method is to direct a stream of gas at low pressure onto the specimen. Air is then introduced to burn up the deposit of carbon. Another method is to provide a nearby cold "finger" on which the organic molecules will preferentially collect. A combination of the two devices may be worthwhile.[2] (See Section 13.8, last paragraph.)

13.7. DEPTH OF FOCUS

Depth of focus of the SEM is the greatest among microscopes. This is because the final condenser lens has such a very small aperture in order to demagnify the image of the electron source sufficiently to make it an adequate probe. Depth of focus is of great advantage for keeping in focus all parts of a rough topography, but a compromise must be made between depth of field and resolution (Figure 13.3). Since the specimen is tilted toward the electron collector, there is a substantial difference in path length from one side of the specimen to the other (Figure 13.1). Most manufacturers have introduced a second lens to correct the focus as the beam moves across the specimen (dynamic focusing).[2,6,7]

13.8. FOCUSING

Focusing the SEM is much like focusing a television set. The operator should first set the contrast and brightness controls to suit his convenience. It is generally easier to focus on a screen that is not very bright. If the contrast is not suitable, one should consider using either the black level or the gamma control. Increasing the black level will make a flat surface appear less flat. Increasing gamma control is useful in looking at the bottom of holes or in getting a picture when the specimen is changing. Both the black level and gamma controls produce artificial pictures, so their use should always be recorded.[7]

For magnifications of over 2000×, the image should be tested for astigmatism with the fine-focusing control by passing through focus. If the picture seems to be smeared at an angle to the scanning axes on one side of focus and at the opposite angle on the other side of focus, there is astigmatism. Correction is by means of the stigmator, going back and forth through focus until correction is optimum.[7]

Focusing and astigmatic adjustments should be checked at a higher magnification than the one to be recorded. If the beam voltage, aperture, or lens currents are changed, the image should be refocused and correction for astigmatism should be made again.[7]

13.9. ILLUMINATION

Illumination in the SEM picture should appear to go toward the top of the specimen. The observer is accustomed to this type of illumination in everyday life and therefore finds it the easiest to interpret in microscopy: Shadows are below elevations and in the upper part of depressions. For single pictures the illumination does appear to come from the top, but with stereopairs the illumination usually appears to come from the side of the integrated image. Effective illumination from the top of the stereopair can be obtained by tilting the specimen about an axis parallel to the line between specimen and detector. Since this kind of tilt is not usually provided in SEM stages, a goniometer stage can be used to rotate the specimen on an axis normal to the specimen-detector line.[2]

The source of illumination in the SEM needs to be an electron gun of great intensity because the specimen's responsive signals are relatively weak. The thermionic gun made of tungsten filament (the same as for TEM; see Figure 12.2) is reasonably satisfactory at moderate magnifications, but for higher magnifications or shorter scanning times, more intensity is needed. The lanthanum hexaboride gun[2] (Figure 13.7) of

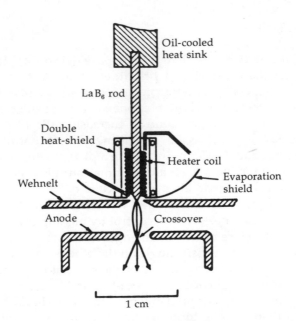

FIGURE 13.7. Electron gun with LaB_6 cathode[2] (from Broers).

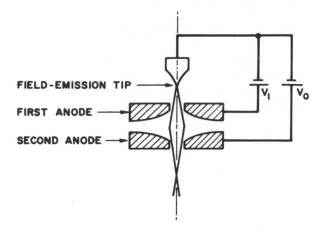

FIGURE 13.8. Electron gun with field emission[2] (adapted from Crewe *et al.*).

Broers is 10 times as bright as the tungsten filament and has a longer life at a cost of additional complication and expense.[5] The field-emission gun[2] (Figure 13.8) is capable of very high intensity and small spot (less than 0.2 μm).[5] The major problems are with the high vacuum required and the relative instability.[2]

13.10. RADIATION

The radiation utilized in the SEM is one or more of eight different types of signal, as indicated in Table 13.1.[6] So far, we have been discussing the emission and reflection from a surface of *secondary* and *backscattered* electrons as signals of information in the SEM. However, some electrons are *transmitted* to give information about the thickness and composition of foils and films. The degree of *absorption* of electrons can thereby be determined so as to give complementary information about the specimen.

In certain specimens, high-energy electrons create pairs of electron holes. If the pairs combine, energy is manifested as radiation of long wavelengths. Such *cathodoluminescence* can include visible light of various wavelengths (colors). Cathodoluminescence is usually weak and requires an efficient light-collecting system, including elliptical mirrors and fiber optics.[2] Results are both qualitative (pictorial) and quantitative (photometric), and are feasible from both industrial and biological specimens.

TABLE 13.1
Types of Radiation Signaling Information in SEM[6]

Radiation	Signal	Location	Information
Emission	Secondary electrons	Within 5 μm of surface	Topography of surface
Reflection	Backscattered electrons	Within 1 or 2 μm of surface	Nature of specimen
Transmission	Transmitted electrons	Thin foils and films	Thickness and composition
Absorption	Specimen current	Through specimen	Complementary to information above
Beam induction	Current in external circuit	Within specimen	Semiconductors
Cathodoluminescence	Photons of selected wavelength	Light	Various phases in a specimen
x-Rays	Selected wavelengths of x-rays	Kinds of atoms	Spectrochemical analysis
Auger	Auger electrons, selected wavelengths	Auger electrons	Chemical elements present

X-Radiation

When a beam of electrons impinges upon a solid, x-ray photons may be emitted by: (1) core scattering, manifested as a *continuous* spectrum of x-rays, and (2) inner-shell ionization, which yields the *characteristic spectrum* of x-rays.[2]

The intensity of the continuum is a function of atomic number and accelerating voltage. As the voltage increases the continuum manifests shorter wavelengths and increased intensity. The extent of the continuum radiation also increases with increasing atomic number, because heavier elements have more nuclear scattering and less loss of energy by interactions between electrons. The continuum radiation forms the background x-radiation[1] in the electron microprobe-scanning electron microscope. In most instances it is desirable to keep the continuum (background x-radiation) to a minimum.[2]

Characteristic x-radiation is by the interaction of incident electrons with the electrons belonging to the inner shells of the atoms in the specimen. If an incident electron has sufficient energy, it may dislodge an electron from an inner shell (e.g., K, L, and M[1] (Figure 13.9[2])), leaving the atom in an excited (ionized) state. The atom returns to the original state by transition of an outer electron into the vacancy in the inner shell, losing specific energy by the emission of a photon of

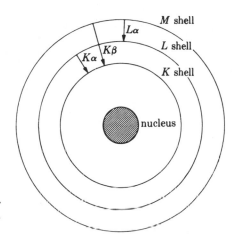

FIGURE 13.9. Scheme of electronic transitions in an atom.[2] Courtesy of J. I. Goldstein, H. Yakowitz, *et al.*

x-radiation (Figure 13.10). The discrete energy level is described by a principal quantum number of the atom: one energy level for $n = 1$ (K shell), three energy levels for $n = 2$ (L shell), five energy levels for $n = 3$ (M shell), etc. The emitted x-ray photon has a discrete energy level equal to the difference in energy between the initial and final states of the atom. That is, the *wavelengths of the characteristic x-ray photons* from a specific chemical element are critically related to a discrete energy level, E, in keV.* Determinative tables are available.[2]

Figure 13.10 indicates that following ejection of an orbital electron (top), electron relaxation (center) may cause an *Auger* electron to be ejected (left, bottom) *instead of* an x-ray photon (right, bottom). In either case, the energy is characteristic of the emitting element. The *Auger yield is very high for light elements,* while the x-ray yield is low, so that Auger analysis may prove to be more useful than x-ray analysis for light elements.[2,13]

---

* The electronic transition involves a change from one discrete energy level to another, and the change in energy ΔE (by Planck's law) is equal to $h\nu$, where ν is the frequency associated with the transition. In terms of wavelength λ, since

$$\lambda\nu = c \quad \text{(the velocity of light)}$$

then

$$\nu = c/\lambda$$

and

$$\Delta E = hc/\lambda$$

where h and c are constants and ΔE can be expressed in kiloelectron volts (keV) instead of ergs (1 keV $= 1.602 \times 10^{-9}$ erg).

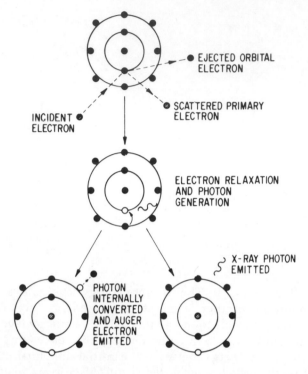

FIGURE 13.10. Scheme of electron excitation manifesting characteristic x-rays or Auger electrons.[2] Courtesy J. I. Goldstein, H. Yakowitz, *et al.*

The qualitative and quantitative use of x-rays excited by a microprobe of electrons has been offered separately by manufacturers of the electron-probe microanalyzer (EPMA). However, the EPMA and SEM can be one and the same instrument,[2] and they are treated as such in this chapter.

If a solid-state *x-ray* detector (SD, Figure 13.1) is placed *close* to the electron probe of an SEM or EPMA, x-rays pass through a thin beryllium window into an evacuated chamber containing a cooled reverse-bias lithium-drifted silicon crystal. Absorption of the x-rays produces a charge pulse which is converted to a voltage pulse by a sensitive preamplifier. After more amplification, a multichannel analyzer sorts the voltage pulses and displays them on a cathode-ray tube, *x–y* recorder, or computer.[2]

13.11. USEFUL MAGNIFICATION

One aspect of the SEM's *useful magnification* is different from TEM's: The micrographs taken on the SEM cannot be enlarged very much because of the raster lines.[9] The limit of directly useful magnification is about 20,000×; higher magnifications tend to be fuzzy.[8]

Useful magnification can be determined on manufactured linear scales of a variety of spacings. A composite photomicrograph should be made of a particular scale in two positions of rotation, 90° apart. A better set of standards is of cross-ruled diffraction gratings. All standards should be photomicrographed at the working distance used to record the image of the specimen.

13.12. FIELD OF VIEW

The *field of view* of the SEM is comparable with that of the light microscope, and is large compared with that of the TEM[7] (Figures 13.2,[3] 13.11, and 13.12[10]).

13.13. NOISE

Noise should not be greater than one-fifth the signal change caused by the specimen, if the eye is to distinguish the true signal change in the

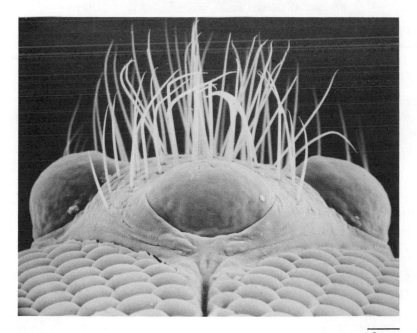

9 µm

FIGURE 13.11.   Eyes and back of head of sweat bee, taken by F. G. Rowe with a Coates and Welter SEM, Model 106A with field-emission electron gun (Figure 13.8).

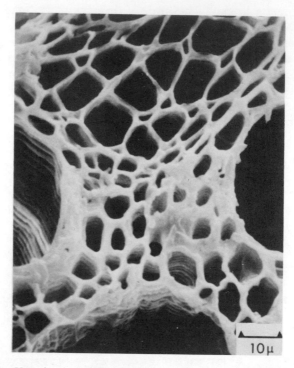

FIGURE 13.12. Vascular bundle in *Scirpus lacustris*, a reed.[10] Courtesy of A. C. Reimschuessel.

scanning-electron image. Noise can be introduced at all stages in the signal path from the specimen to the cathode-ray tube (Figure 13.1). In the transmission electron scintillator–photomultiplier system, the number of informative electrons or photons is smallest at the stage of intersection between beam and specimen, where the secondary electrons are collected. In an efficient scintillator the signal is increased preferentially, but without further information in the image. Noise manifests itself as graininess ("snow") in the image. It results from too little current for the scanning beam, which should be at least twice the threshold (minimum acceptable) current.[2]

13.14. CUES TO DEPTH

The *cues to depth* in the SEM image are very impressive and effective. With single SEMs the main cues are the distinct shadow and the apparent perspective. The microscopic specimen appears in the SEM image as though one were looking at macroscopic elevations and depressions of wet sand (Figure 1.4). In both cases, the depressions have their shadows at their lower halves of the visual image. The illustrator must assume the responsibility for correctly orienting photocopies of his photographs in his reports and publications. The shadows should fall in the "natural" azimuth. Perspective should also be "natural." Parallel lines should appear to converge; larger parts should be in front of smaller ones. Note these aspects in Figures 13.3, 13.5, and 13.11.

13.15. WORKING DISTANCE

The long *working distance* of the SEM is of special advantage in arranging for any movement of the specimen. This is the distance between the surface of the specimen and the front surface of the objective lens.[1] Working distance from the bottom pole piece of the objective lens to the specimen's surface is around 5–25 mm,[2] so that the specimen and its low-energy secondary electrons are outside the magnetic field of the lens. The working distance also makes it possible to have a large space to hold the specimen upon a special stage for heating, cooling, bending, ion-etching, or electromigration.[9]

13.16. DEPTH OF FIELD

The great *depth of field*[1] is one of the most important advantages of the SEM (Figure 13.2).[2] It is hundreds of times greater than in the light microscope of the same magnification,[5] and is capable of showing an inclination with its highest and lowest points in simultaneous focus. Another need for great depth of field in focus is set by the tilted specimen with consequent differences in distance traveled by the electrons.

13.17. STRUCTURE

Structures are composed of units of increasing complexity: atoms, molecules, domains, grains, and cells.[1] Periodic structures are particularly interesting at this point. Figures 13.13 and 13.14, for example, show the *three*-dimensional periodicity of certain diatoms.[10] Figures 13.15–13.18 show the linear periodicity, respectively, of yellow, pink, orange, and black scales taken from colored butterflies' wings.[10] The same figures also show interlinear structure which may or may not contribute to the structural color.[11]

20 μm

FIGURE 13.13. Filter aid: diatomaceous earth. Taken by F. G. Rowe with a Coates and Welter SEM, Model 106A with field-emission electron gun (Figure 13.8).

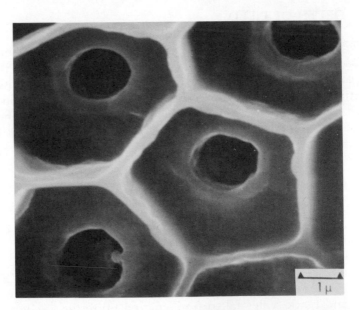

1 μ

FIGURE 13.14. A single fragment of shell in diatomaceous earth.[10] Courtesy of A. C. Reimschuessel.

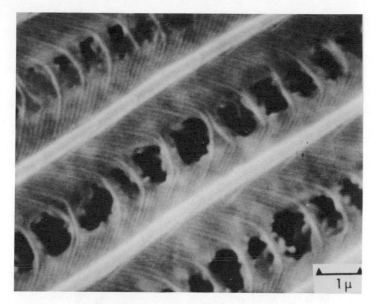

FIGURE 13.15. *Yellow* scale of wing of the butterfly *Colias eurythyme*. The periodicity is 3–4 μm.[10] Courtesy of A. C. Reimschuessel.

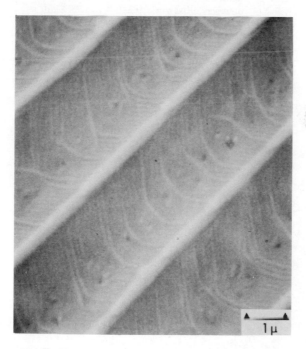

FIGURE 13.16. *Pink* scale of wing of the butterfly *Colias eurythyme*. The periodicity is 2–4 μm.[10] Courtesy of A. C. Reimschuessel.

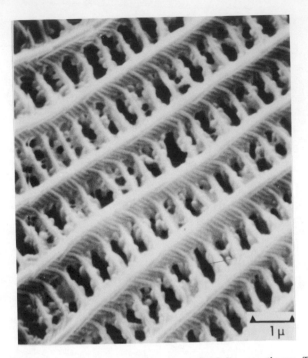

FIGURE 13.17. *Orange* scale of wing of the butterfly *Colias eurythyme*. The periodicity between ribs is 1.5 μm; between cross-ribs, 0.5–0.7 μm.[10] Courtesy of A. C. Reimschuessel.

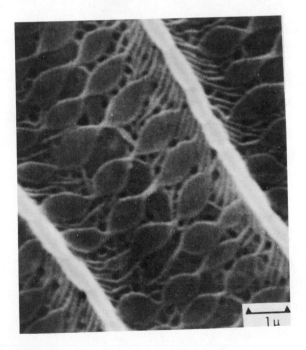

FIGURE 13.18. *Black* scale of wing of the butterfly *Colias eurythyme*. The periodicity is 3–6 μm.[10] Courtesy of A. C. Reimschuessel.

13.18. MORPHOLOGY

Morphology, as has been stated,[1] involves the size and shape of a structure. In Figures 13.15–13.18, the distance between lines is to be compared with the wavelength (color) of light from the butterfly's wings.

13.19. INFORMATION

To continue our search for *information* about the color of the scales from butterflies' wings it is necessary to know macroscopically whether the color is yellow, pink, orange, or black. Is a pigment present to contribute to the total color by reflected light? What is the color by transmitted light with the specimen mounted in a permeating liquid of similar refractive index, as compared with the color in a liquid of very different refractive index?[11] With such information, scanning electron micrographs such as Figures 13.15–13.18 have much more meaning than would be possible otherwise.

13.20. DYNAMIC EXPERIMENTATION

Dynamic experimentation is a specialty with the SEM, since it has such great depth of field in focus, coupled with a large working space. Substantial specimens may be moved with considerable freedom, with or without accessories.

SEM stages strong enough to stretch metals, alloys, and semiconductors have been devised.[2] Micromanipulators have been built to operate on specimens in the SEM. Closely related mechanisms are made for observing and measuring scratch and indentation hardnesses.[9]

Special stages also have been developed to change the temperature, magnetic field, electric field, liquid medium, etc.[2] Heating, cooling, mechanical manipulation, etching, and electromigration can all be performed on the specimen within the SEM. For example, explosive chemicals have been studied as they slowly decompose at carefully controlled temperatures below 200°C. Nickel particles have been thermally etched at 1000°C or 1300°C. Change in topography of the surfaces of metals has been studied at various temperatures within the SEM.[9]

Ion-etching has been performed by an argon ion beam at 5 keV striking the specimen at right angles. Some of the materials studied *in situ* by ion-etching are dental tissues, iron oxide films, aluminum on oxidized silicon,[9] and a wide range of inorganic and organic substances.[7] Electromigration is a process in which metal atoms are moved along a conductor while a current is flowing in it. Aluminum has been studied in this way.[9]

With beam diameter greater than about 1000 Å the current can be made strong enough to employ television rates of scanning. Then dynamic events may be recorded either on cinefilm or on video tape. With narrower beams the SEM is limited to imaging repetitive events which can be examined stroboscopically.[9]

13.21. BEHAVIOR OF THE SPECIMEN

Behavior of the specimen can be observed "postmortem" by exposing the specimen to conditions outside the TEM's chamber, stopping the reaction, and putting the spent specimen into the TEM. Such procedures include weathering, wear, fracture (Figure 13.19), corrosion, erosion, annealing (Figures 13.4 and 13.20), and certain manufacturing processes.

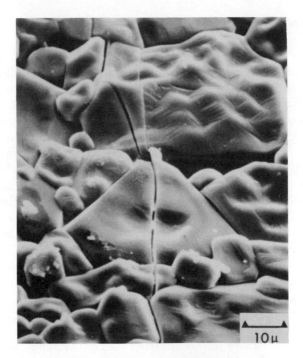

FIGURE 13.19.   Intragranular fracture in alumina.[10] Courtesy of A. C. Reimschuessel.

FIGURE 13.20.   Pure gold wire before annealing. Compare with Figure 13.4, same sample after annealing.[10] Courtesy of A. C. Reimschuessel.

13.22.   PREPARATION OF THE SPECIMEN

The *preparation* of a specimen for examination in the SEM by secondary electrons can range from almost nothing to a complex procedure. No preparation is needed if the surface as received is of interest and is conductive. If the interesting surface is an insulator, it can charge up and ruin the image. If the insulating layer is thin enough for some electrons to penetrate onto an underlying conductor, a satisfactory image may be obtained from the secondary electrons. If not, the usual procedure is to coat the surface by vaporizing with a thin layer (~ 100 Å) of metal such as gold or gold plus palladium. A different approach is to view the insulating surface through a hole in a foil of aluminum.

The surfaces of wet specimens are preserved by freeze-drying. For example, the fibers of wet paper fresh from the machine are not collapsed during freeze-drying, whereas they are by ordinary drying. Of course, in finished paper the fibers are not only flattened by the commercial drying but even more so by the calendering (glazing) process.

FIGURE 13.21. Section of *Scirpus lacustris*, a reed, to show distribution of stomata.[10] Courtesy of A. C. Reimschuessel.

The freeze-dried surface does not represent the surface of paper, but it helps us to understand the changes in the fibers during paper-making.[2]

In many instances, a suitable surface of the material has to be made for the SEM. Biological material may be cut once with a razor blade or microtome knife (Figures 13.12 and 13.21). Watery tissue may be frozen and fractured. The size of the ice crystals might be controlled by adding glycerine or gelatin or both. Or the water can be displaced by alcohol or acetone, the sample quenched in liquid nitrogen, and then fractured.[9] Treatment with hydrophobic liquids, such as chloroform, ether, amyl acetate, or epoxy resin (unpolymerized) may change the appearance of the material so much that the image may not be interpretable. More complicated procedures, such as chemical digestion, may make interpretation easier.

13.23. PHOTOMICROGRAPHY

Photomicrography is simpler with the SEM than with the TEM because the image is exposed to the photosensitive material outside rather than inside a vacuum. With the SEM, Polaroid® or other fast-developing photosensitive material may be used directly in the commercial camera back, so that photomicrographs may be made very quickly.

On the other hand, photomicrographic exposure is longer with the SEM than with the TEM. SEM signals are integrated slowly into the image, since less noise (fewer dots) appears during slow integration than during fast exposure. An exposure of 40 sec is usual; 100 or 200 sec may be required to clear up the noise. Yet a mere 20-sec exposure will record a transient image and prevent damage to the specimen by the electron beam. The number of lines per frame should be adjusted to 1000. More lines per frame could overlap and blur the picture, while fewer lines could be so far apart as to show in the final photomicrograph.[7] In any event, an SE micrograph with its raster structure cannot be enlarged to the same extent as a TE micrograph with its very fine grain structure.

Many SEMs have cameras which automatically select exposure times according to overall brightness and contrast in the image. If the operator is interested in optimum brightness and contrast for a small part of the image, he can override the automatic mechanism with a manual control. With instruments having no automatic cameras the operator has to make his judgment by studying a line across an interesting part of the image in darkness. The scan line should not be left very bright on the record screen because the screen is easily burned, and is expensive to replace. Using Polaroid® positive film matched to the ASA rating of the negative film speeds up the trials with fewer errors.[1] Scan rotation units make it easier to arrange the image in a position that lends itself more readily to interpretation.

13.24. SUMMARY

The scanning electron microscope (SEM) differs from the transmission electron microscope (TEM) in several important respects:

1. SEM deals with various signals *scattered* from the *surface* of the specimen.

2. These signals are received by *cathode-ray tubes* which are *synchronized with the scanner* probing the surface of the specimen.

3. The picture (pattern) on a cathode-ray tube, like that on a television tube, is composed of scanner (raster) *lines* which limit the degree of useful total magnification, but permit photomicrography with a camera placed outside any vacuum. The Polaroid® type of quick photography is the most popular.

4. The photomultiplier receives the secondary and/or backscattered electrons at an *average* angle to the surface of the

specimen and at *various* angles to various facets of grains. The resultant image manifests modulated *shadows* and a perspective which add three-dimensional aspects and render the image *easily interpreted.*

5.   A faceted or rough surface manifests *topographic contrast* because the detector is highly directional. Only certain facets are so oriented that most of their backscattered electrons are aimed exactly at the detector, while the remaining facets contribute fewer electrons so as to make less contrast.

6.   Crystalline specimens also contribute contrast by electron channeling and by magnetic or semiconducting effects.

7.   In addition, there are noncrystalline specimens contributing contrast by other effects of the initial electron beam: magnetic, voltage (from beam-induced current), and luminescence.

Like the TEM, the SEM has *great depth* of field in focus because the wavelength is so short and the angular aperture is so small. The resolution by means of the SEM is somewhat less than that of the TEM, but is, of course, greater than the resolution of the light microscope.

The most important comparison, however, emphasizes the *differences in the kinds and extent of information* provided by the three different types of microscopes.

8.   The qualitative and quantitative use of x-rays excited by a microprobe of electrons is offered separately by some manufacturers as the electron-probe microanalyzer (EPMA). Other manufacturers offer the EPMA together with the SEM as one instrument.

Field-Emission Microscopes

14.1. INTRODUCTION

The two kinds of field-emission (point-projection) microscopes*
were both invented by E. W. Müller[1] (Figure 14.1). The older device
(1936) employs field-emission of *electrons* from the *negatively* charged
tip of a very sharp needle in a vacuum by point-projection of the image
onto a positively charged, fluorescent screen. The device is a specialized
cathode-ray tube (Figure 14.2).[2] The newer device (1950) emits *ions*
from an *anode*.

The electron field-emission microscope† is unique in its ability to
detect, if not resolve, single atoms or small groups of atoms on the
surface of the needle tip. Thus the purity of the metal is established. If
the coil shown in Figure 14.2 is coated with a potential additive such as
barium, and heated, such atoms will migrate to the metallic tip of the
cathode and be visible as bright spots in motion. The device is useful in
studying electron emission as a function of surface structure. Organic
compounds can be substituted for the inorganic additive and evaporated onto a clean metal point. Phthalocyanine shows a pattern that is
curiously like the structural formula of its molecule.[2] Caution must be
taken, however, in the interpretation of such patterns since a large

---

* Field-emission (point-projection) microscopes are usually operated with the
  projection-point *cold*. In the electron-emission microscope the specimen can be heated
  (Figure 14.2). *But* there are other kinds of *hot*-emission microscopes, in which the
  specimen may be *broad* and may glow with *visible* or *infrared* radiation (becoming the
  source of illumination). If, however, the surface of a broad *hot* specimen is *irradiated* with
  ultraviolet light, photoelectrons may be extracted at high potential and imaged on a
  fluorescent screen.[3]

† The field electron microscope is sometimes loosely called a shadow microscope. It is not
  to be confused with another, very different shadow microscope,[5] much like the TEM
  except that it produces an image by direct (point) projection instead of by a projection
  lens.[2]

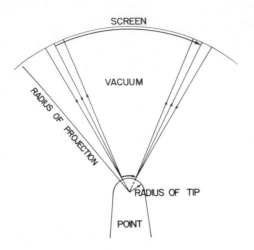

FIGURE 14.1. Principle of both kinds of point-projection or field-emission microscopes.[1]

number of molecules, regardless of their shape or symmetry, produce similar patterns.[4]

Müller's newer device (1950) is the field-ion microscope.[1,6] It operates with the metallic needle tip *positively* charged in the high vacuum. A gas, such as helium, is introduced and is ionized at the tip of the cooled specimen. The tip is needled down from fine metal wire (e.g., tungsten) by etching to give it a radius of only 5–100 nm. The etching roughens the tip of the metal so that its atoms stand out in a terrace of ledges (Figure 14.3).[3] A potential of several kilovolts is applied to the tip, positive with respect to the negatively charged fluorescing screen. Figure 14.4 explains the formation of an image (or pattern) on the

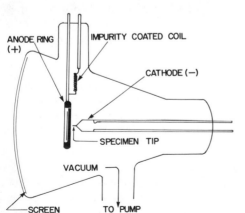

FIGURE 14.2. A schematic drawing of the electron field-emission microscope, showing the remarkable simplicity of the apparatus. The coil coated with an impurity may be heated to evaporate its atoms onto the point for the purpose of making emission studies.[1] Courtesy of E. W. Müller.

FIGURE 14.3. Illustration of a ball model which approximates to the end of a field-ion emitter. The surface is not smooth; protruding atoms are imaged in the field-ion microscope.[3]

fluorescent screen.[1] Incident atoms of the gas (e.g., helium) are attracted to the rough, sharp tip of the specimen. Some of them hop along the tip, losing electrons to the specimen cooled by a cryogenic bath. The resulting ions of gas are projected onto the fluorescent screen, where they release light in a typical pattern (micrograph) (see Frontispiece).

Figure 14.5[3] is a diagram of the typical apparatus. Field-ion microscopy is concerned with the study of the atom itself, not only in metallurgy, but also in fundamental physics and chemistry.[7–10]

Professor Müller added the *atom probe* to the field-ion microscope so as to identify atoms chemically. Selected atoms are fed into an ion detector. The data are collected slowly by a skilled operator, and require careful interpretation.[3,9]

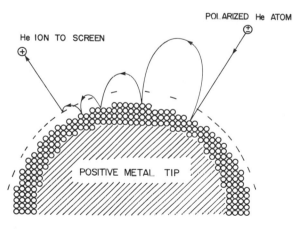

FIGURE 14.4. A polarized helium atom is attracted to the metal tip, is slowed down in a number of hops, and is ionized in the ionization zone (0.2 Å thick) over a protruding atom. The helium ion is accelerated toward the screen.[1]

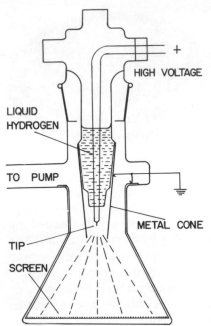

FIGURE 14.5. Schematic drawing of a field-ion microscope.[1]

14.2. ATTRIBUTES CONTRIBUTING TO VISIBILITY BY FIELD-EMISSION MICROSCOPY

14.2.1. Thought, Memory, and Imagination

The mental processes—*thought, memory,* and *imagination*—are especially important in the interpretation of images by field-emission microscopy. In the absence of lenses, illuminating devices, and automatic controls, a field-emission microscope is less of an apparatus and more of a direct aid to the eye and brain. Together the mental processes furnish, Müller wrote,[1]

> the skill, experience, patience, and ingenuity which the microscopist can muster for the preparation of the specimen, the making of the observation, and, finally, the interpretation of the image. It seems as if the evasive atoms still hide from the curious eye of the casual sightseer and reveal themselves rewardingly only to the serious researcher.

14.2.2. Resolving Power

The field-ion microscope (FIM) is capable of visually separating individual atoms.[1–3,7] Each atom is depicted as a bright dot on a black background, like stars in the night sky. The theoretical resolution is about 1.5 Å, since the wavelength of ions (protons) is so much less than that of electrons, and since the temperature of liquid hydrogen or nitrogen is used to reduce thermal motion.[2]

14.2.3. Resolution

The practical *resolution* of the modern FIM is as small as the triangular spacing of 2.3 Å. The resolution amounts to the separation of bright spots or halos. This resolution is with certain elements, with helium ions, and with liquid hydrogen. Liquid-nitrogen temperature is almost as good. Neon, nitrogen, oxygen, and argon may be used instead of helium, but the images are inferior. The elemental specimens which give best resolution and those which give satisfactory images will be discussed in the section on the structure of the specimen,[1,9] Section 14.2.15.

It should be emphasized that the resolution with the FIM is the separation of atoms as halos (bright spots or tiny stars). There is no resolution within the atom. In the case of the field-emission electron microscope, while the resolution of phthalocyanine (Figure 14.3) is 12 Å on a side,[2] each square is divided into four or two parts, giving a resolution of at least 6 Å.

14.2.4. Contrast

Usually there is plenty of black and white *contrast* between the bright spots. In 1960, Müller used color contrast to detect quick changes in two successive images by printing one on red film and the other on green. When the two are superposed, deletions from the first pattern are red, additions green, and unchanged portions yellow.[4,9]

The frequent problem of low intensity of field-ion images may be termed one of contrast. Müller found some success by post-accelerating the ion beams through a fine, high-transmission wire mesh, but disturbing double images may occur. Conversion of the ion-image into an electron image by using secondary electron emission from a fine grid is somewhat more promising. External electronic image intensifiers based on photoelectric emission have proved to be more successful.[1]

14.2.5. Aberrations

Aberrations in field-emission microscopy refer to those originating either at the tip-emitter or at the fluorescing screen (since there are no lenses). The tiny tip is the chief source of aberrations. Most of them arise from the mechanical stress exerted by the electric field on the conductive surface, about 1 ton/nm², which may exceed the yield strength of the particular metal by a factor of 50. The shear component of the stress, due to nonsphericity of the tip, may cause dislocations to move until the tip fractures. One correction is to reduce the tip's radius to less than 200 Å, so that there is insufficient room for defects to grow.[1]

Another kind of correction is to reduce the field strength in order to shape the tip without introducing severe lattice imperfections. Hydrogen promotes field evaporation and therefore shaping of the tip; it can reduce the required field strength by 5–20% for most metals as tips.

14.2.6. Cleanliness

Cleanliness is of supreme importance in field-ion microscopy because some of its main applications involve the purity of metals, the detection and location of impurities, and the study of solids, liquids, or gases adsorbed on the surface. Therefore, the specimen of metal used for making the emitting tip should be as pure as possible. So also must be the gases for ionization or for adsorption, and any other substance to be introduced into the microscope. All these ingredients must also be scrupulously clean of dust, dirt, oxidation, and any other contamination. The metal for the specimen tip should be kept free of mechanical changes that would alter crystallographic conditions, such as dislocations, slippage, twinning, orientation, etc.

14.2.7. Depth of Focus

Depth of focus does not apply to the microscope itself since it is without lenses, but it does apply to the photographic lenses in the cameras used to record either snapshots or moving pictures. With the field-ion microscope, especially, the images are of such low intensity that the lens should be of high aperture (low *f*-number). This means short depth of focus, which in turn means that the fluorescent screen should be as flat as possible.[4] Another reason for a flat screen in either kind of emission microscope is that images often are changing or migrant. This means short exposures, wide-open lens, and consequently short depth of focus.

Generally the atoms are in *focus*, but occasionally an atom will jump out of focus and appear brighter on the fluorescent screen.[9]

14.2.8. Illumination

Illumination in emission microscopy is by the point-projection of electrons or ions from the specimen-emitter onto the fluorescent screen. Intensity of image in the field-ion microscope depends on the specimen,

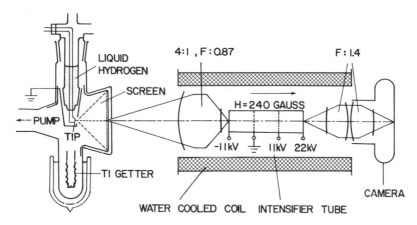

FIGURE 14.6. Schematic diagram of a field-ion microscope with external image intensifier.[1]

its tip, the ionizable gas, etc. When these factors cannot be improved upon sufficiently for visual examination or for photography, Müller has used an external image intensifier (Figure 14.6).[1]

14.2.9. Radiation

The *radiation* in the field-emission electron microscope is a cone of electrons, as in an ordinary cathode-ray tube. In the field-ion micro-scope, a stream of polarized atoms of a gas such as helium is attracted to the cold specimen at high positive potential. Here the gas atoms are slowed down in a number of hops, and are ionized and accelerated toward the screen.[1,3] The imaging process is not yet fully understood.[7]

Most of the metals examined to date are isometric and thereby isotropic. Some, however, such as rhenium, are in the hexagonal system and therefore *anisotropic*. So far, anisotropy has not contributed to visibility in the field-ion microscope.

Orientation of the metal grains, however, does contribute to the field-emission image, so that grain boundaries, twins, dislocations, etc., are recognized.[1]

In the field-electron microscope, the orientation of some molecules, such as phthalocyanine, apparently makes a difference in the image, but the explanation is not clear.[4]

14.2.10. Magnification

Magnification is quite incidental in field-emission microscopes. It is primarily dependent on the radius of curvature of the tip. But the variation in local magnification due to bumps on the tip is not directly related to the radius of the bump because of a strong compression factor.[4] Magnification, of course, also depends upon the projection distance.

14.2.11. Field of View

The *field of view* in emission microscopes is small because the tip has a volume of only about 10^{-21} m³. Larger fields would need higher potentials than the usual 30 kV.[3] Larger volumes would also result in higher stresses on the specimen with greater potential damage.

14.2.12. Artifacts

The chief *artifacts* in field-ion microscopy come from disturbance of the specimen by the extremely high mechanical stresses induced by the very high electrical fields. Such stresses are great enough to cause fracture. Stress both nucleates and rearranges dislocations. It can induce deformation twinning and usually leads to relatively high artifactual vacancies. So long as such artifacts are understood they can be turned into an advantage in performing experiments *in situ*.[3]

In the field-ion microscope (FIM) the *third dimension* is limited to one or two layers of atoms, a true surface. Yet in a homogeneous specimen we can infer much about the third dimension from the surfacial arrangement of the atoms and from previous knowledge of the space-lattice of the respective metal (Figure 14.3).[3] Already field-ion microscopy is accepted in the study of three-dimensional metallography.[11]

14.2.13. Working Distance

The *working distance* in both kinds of field-emission microscopes is fixed by the dimensions within the sealed chamber. Nevertheless, the working distance is ample for any practicable experiment, since no lenses are involved.

14.2.14. Depth of Field

The *depth of field* in emission microscopes is completely adequate because there are no lenses (and no focusing), and the object space consists of only one or two rows of atoms.

14.2.15. Structure

The atomic and crystalline *structures* of the metal are very important attributes contributing to visibility in the field-ion microscope. Also important is the chemical nature of the ionizable gas. By the ionization of helium, for example, the following metals at the temperature of liquid hydrogen yield satisfactory field-ion images: tungsten, rhenium, iridium, platinum, molybdenum, tantalum, niobium, and rhodium. Marginal are: zirconium, vanadium, palladium, titanium, nickel, and iron. With neon gas gold, iron, nickel, copper, and zinc can also be imaged at or below their evaporation rate, as well as the metals in the first of the two lists above. Various changes in crystalline structures can be seen at the atomic level: dislocations, vacancies, impurity atoms, twinning, slip bands, and fatigue cracks.

14.2.16. Morphology

Morphology of the field-emission microscopical specimen at present refers chiefly to the sizes and shapes of the particles of a second phase distributed in an alloy. Quantitative particle-size distribution is difficult to obtain due to the complex shape of the specimen. Calculation is best performed on a computer.[3]

14.2.17. Information

Crystallographic *information about the metallic specimen* amenable to field-ion microscopy is published and generally available.[8-10]

14.2.18. Experiments

Various *experiments* may be performed inside the field-ion microscope. By preferred field evaporation of weakly bound, substitutional impurity atoms or by corrosion through field-induced chemical reactions with strongly adsorbed oxygen, carbon monoxide, or nitrogen, surface vacancies may be made to appear. Slip bands can be produced *in situ* by pulling at the tip through application of a reduced electric field at elevated temperature. The growth of fatigue cracks has been seen by superimposing an alternating-voltage component onto the tip-voltage in order to cycle the stress. The difference between annealed and field-evaporated specimen tips can be watched. The reversed action—rearrangement of atoms during annealing of a field-roughened tip—can be witnessed and, at the same time, activation energies can be measured *in situ*.[1] These are but a few examples of experimentation in the field-ion microscope.

Experiments may also be performed by field-emission electron microscopy. Emission is critically influenced by monatomic layers, and single atoms or small groups of atoms can be detected as scintillating bright spots on the fluorescing screen. Migration of the atoms is readily followed as the surface structure is varied. Organic molecules may be introduced by evaporation from the heater-coil within the microscope. The scintillating spots have structure, but interpretation awaits further experimentation.[2,4]

14.2.19. Behavior

Behavior may be observed in field-emission microscopes, since the environment may be changed relatively easily in such a small, sealed space. Oxygen, carbon monoxide, nitrogen, and other atmospheric gases may be introduced at will. Likewise, various other molecules may be introduced if they can be volatilized satisfactorily within the micro-scopical chamber by means of the high vacuum and localized heat. In this way corrosion and stress-corrosion may be studied *in situ*.[1] Behavior of the specimen in organic atmospheres may also be studied.[2]

14.2.20. Preparation of the Specimen

Preparation of the specimen is a most important attribute contribut-ing to imaging by radial projection. The first crude steps are the chemi-cal etching and electropolishing of a fine wire of the specimen metal, such as platinum. The etched specimen tip is spot-welded to the tungsten leads sealed in glass (Figure 14.5). The microscope is evacuated, and evaporation occurs at the tip when thermal activation permits atoms to leave the surface. However, when a sufficiently high electric field is applied, some surface atoms may leave as ions by shed-ding an electron or two apiece. Field evaporation takes place preferen-tially at the sharp tip which had been roughened by the etching. The tip becomes an oblate microhemispheroid, characterized by a uniform field-strength.[1]

The problems of preparing *photographs* of field-emission images are those of recording a fleeting or moving subject at a low level of light intensity. A fast photosensitive emulsion is required. Fortunately, there is plenty of contrast: white spots on a black background. Visual monitoring may be made in a darkened room. The external image inten-sifier shown in Figure 14.6 is helpful in severe cases, especially in taking moving pictures.[1,9]

Even with fast film, fast lenses are required in field-ion microscopy. Hydrogen-ion still pictures can be taken at $f = 1.4$, but with helium ions $f = 1.0$ is required. For 16-mm motion pictures $f = 1.0$ is also re-quired.[4,9]

Müller reports the trick of taking two successive exposures and superimposing them to make a positive picture of very fast changes.

The neatest part of the trick is to use color contrast. Print the first picture on red film and the second on green.[1,9]

14.3. SUMMARY

A *field-emission microscope* (FEM) is one in which the specimen is the source of radiation by virtue of a strong electrical field and a very high vacuum. The specimen source is cold (at or below room temperature) in contradistinction to thermal emission. Lacking lenses, cold field-emission microscopes achieve magnification by straight projection of the rays of electrons or ions. Therefore the point of the thin-wire specimen is made extremely sharp (5–100 nm radius) With a projection distance of 10 cm, the magnification (over one million times) is useful to the extent of resolving atoms or small groups of atoms, depending on whether ions or electrons are emitted.

The *electron* field-emission microscope is much like an ordinary cathode-ray tube except that the specimen is the cathode and is pointed to a resolving power of 2 nm. In this state the field electron microscope is useful for studying the emission of electrons as a function of surface structure. Impurities can be put into a basket heater to the side of the pointed cathode. In the same way organic compounds may be introduced and studied.

The *ion* field-emission microscope employs the extremely pointed specimen as the *anode*, which is kept very cold in a liquified gas. Helium atoms are introduced and find their way to the cold tip, where they are slowed down in a series of lower and shorter hops on the roughly etched tip of the anode. Some of the slow helium atoms lose an electron to a protruding metal atom. The resultant He^+ ions are shot onto the luminescent screen, producing a dim pattern. The pattern may be photographed on fast film during long exposure or the dim pattern may be intensified photoelectronically. The resultant pattern (image) is refractory metal atoms as they are organized or disorganized on a surface or succession of surfaces. Also made visible are grain boundaries, slip planes, dislocations, interstices, and vacancies. Interpretations lead to surfacial binding energies and work functions. Surfacial chemical reactions of interest include corrosion and catalysis.

X-Ray Microscopy

15.1. X-RAYS

X-rays are electromagnetic radiation of the same nature as visible light but with only 1/1000th the wavelength.[1] X-rays behave like light toward photosensitive materials. For example, when a dentist takes a radiograph he places an unexposed photosensitive film next to the teeth and turns on the x-rays. Cavities show up dark on the negative film because they transmit x-rays so well. Sound teeth show up less dark because they are denser than air or flesh. Ceramic and plastic fillings are still darker if they are denser than teeth and bone. Silver and gold show up very dark when exposed to x-rays because they are very dense (of high atomic number).

The dentist's radiograph is approximately life-size (1×). Dioptric magnification by transmission of x-rays is not feasible because lenses have not been developed for it. X-rays at glancing incidence have been reflected from concave mirrors or from curved surfaces of crystals, but so far focusing is not sharp enough to construct an x-ray microscope this way.

The dentist's radiographs and other radiographs may be usefully enlarged to 10×, but not much more. An x-ray microscope based on the principle of Figure 15.1 has no more magnification than the simple light microscope or equivalent that enlarges the contact negative. The enlarger cannot take out the contact negative's blur, which is caused by the crossover of the peripheral x-rays from such a large source.[2,3] Still, the ordinary source of x-rays may be useful in low-power microradiographic studies of the distribution of heavy elements among lighter ones in an essentially inorganic specimen. In an organic system, contrast in the preferential absorption of x-rays is increased if a selective stain can be found bearing a heavy element in its composition. In any case, it helps in simple microradiography to have the specimen as thin as is

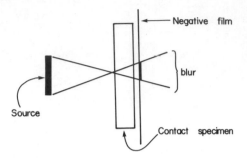

FIGURE 15.1. Ordinary source of x-rays for microradiography blurs the contact image.[2]

commensurate with selective absorption among the constituents of the specimen.

The boom to moderately high-power (100 or 200×) microradiography came with the development of a *point source of x-rays.* As shown in Figure 15.2, the point source provides a very narrow bundle of almost parallel x-rays to impinge upon the specimen. As the source point becomes smaller and smaller, the crossover angle of the peripheral x-rays through the specimen is reduced accordingly. Therefore the contact image is made sharper and the permissible thickness of the specimen may be a little greater.

15.2. CONDENSER LENSES

One way to produce a concentrated point of x-rays is to use the objective lens of an electron microscope to focus electrons on a target situated near the specimen and photosensitive film. Figure 15.3[4] shows

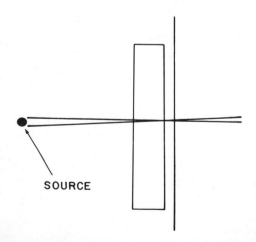

FIGURE 15.2. Point source of x-rays for microradiography improves the contact image.[2]

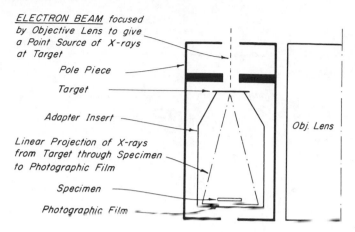

ELECTRON BEAM focused
by Objective Lens to give
a Point Source of X-rays
at Target

Pole Piece

Target

Adapter Insert

Linear Projection of X-rays
from Target through Specimen
to Photographic Film

Specimen

Photographic Film

Obj. Lens

FIGURE 15.3 Schematic drawing of adapter showing target, specimen, pole piece, and paths of electron and x-ray beams.[4]

diagrammatically a homemade adapter for the point-projection of x-rays from a target through the specimen to the film. Figure 15.4 is a line drawing of the adapter which fits into the wide-angle pole piece.[4]

Subsequently, electromagnetic lenses were built into commercial point-projection x-ray microscopes.[2,3,5] Figure 15.5 is a diagram of a

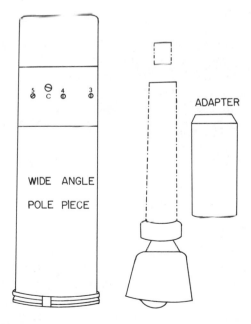

ADAPTER

WIDE ANGLE

POLE PIECE

FIGURE 15.4. Wide-angle pole piece with adapter disassembled.[4]

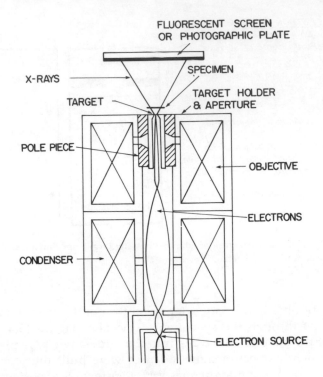

FIGURE 15.5. Diagram of the point-projection x-ray microscope.[3]

double electromagnetic x-ray microscope.[3] Such a microscope is ideal for studying inorganic chemical constituents, especially those of high atomic number. Nevertheless, the specimen should be thin. If it is thick, like catalytic pellets for absorbing automotive leaded gasoline vapors, the pellets are abraded until they are only 25 μm thick. Figure 15.6[6] is a negative microradiograph of thinned unused experimental pellets, taken on a Philips CMR instrument.[5] The photosensitive film was Kodak No. 649 fine-grain, processed through fine-grain developer. The finished negative was enlarged 20×. Other pellets which had been exposed to fumes from leaded gasoline were treated and x-radiographed in the same way. The commensurate negative enlargement is shown in Figure 15.7. The white peripheral rings show the distribution of the lead atoms in the exposed pellets.[6]

Point-projection radiography is also very useful in studying organic systems.[3,7–9] Botty and his colleagues studied plastic foams by employing "soft" (low-voltage) x-rays of low penetrating power, in which low-mass materials (e.g., air and polymers) are differentiated. In Figure 15.8, an enlarged negative microradiograph of an organic artificial foam,

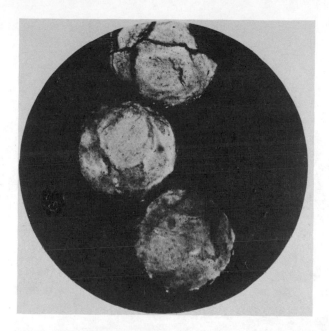

FIGURE 15.6. Microradiograph of experimental auto exhaust catalyst pellets before exposure to leaded gasoline vapors. Thin ground sections (approximately 25 μm thick).[6] Philips CMR Instrument.[5] Pellet diameter: 1.5 mm.

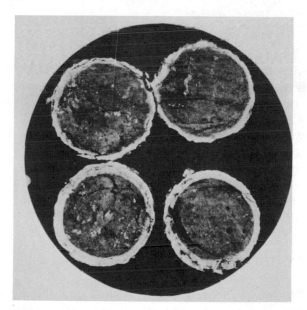

FIGURE 15.7. Microradiograph of experimental auto exhaust catalyst pellets after exposure to leaded gasoline vapors. Ground thin sections. The white peripheral rings show the distribution of Pb condensate.[6] Philips CMR Instrument.[5] Pellet diameter: 1.5 mm.

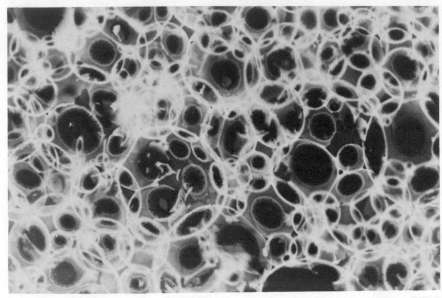

100 μm

FIGURE 15.8. Contact microradiograph of a thin section of thermosetting polymeric foam showing holes, or thin-walled air bubbles (black spots), at the periphery of the thin cell-windows.[7] Exposure conditions: 2 kV, 3.5 min. Philips[5] CMR Unit: standard specimen holder. Kodak No. 649 film.

the *air* bubbles are shown black and the polymeric walls are shown in white and grays.[7] The point of x-rays is so sharp and the specimen is so thin that a useful magnification of 100× was achieved. In fact, contact radiography is useful at 200×, as illustrated by Figures 15.9 and 15.10, showing low-strength and high-strength paper, respectively.[2]

Reflecting x-rays to focus them from *concave mirrors* is a microscopical project of good potential. The theoretical limit of resolution is about 10 nm. Progress has been plagued by the geometrical aberrations of mirror lenses. Therefore, only small apertures of illumination have been used, and practical resolution has been limited to 0.5 μm.[10]

15.3. X-RAY HOLOGRAPHY

X-ray holography holds promise of utilizing the high resolving power inherent in x-rays of short wavelengths. The plan is to make a hologram of *coherent* x-rays. That is, the x-rays scattered or reflected by

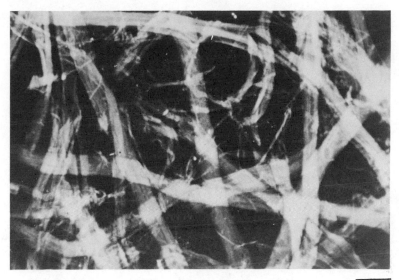

50 μm

FIGURE 15.9. Microradiograph of low tensile strength paper sheet.[6] Philips CMR Instrument.[5]

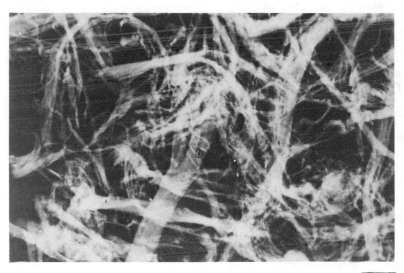

50 μm

FIGURE 15.10. Microradiograph of high tensile strength paper sheet.[6] Philips CMR Instrument.[5] The high strength is attributed to the intermeshing of fibrils with (larger) fibers. Compare with Figure 15.9.

the object are recorded on a photographic plate which is simultaneously illuminated with coherent x-rays from the same source to form an interference pattern. The resultant hologram is to be reconstructed by coherent light into an x-ray micrograph. The basic problem is the development of a coherent source of x-ray illumination.[10]

15.4. SUMMARY

X-rays hold great microscopical promise for several reasons:
1.   Common wavelengths are short enough (0.1–1 nm) to give high resolving power.
2.   Absorption of x-rays is spread among the chemical elements as the cube of the atomic number.
3.   Since x-rays travel well in air, making vacuum unnecessary, wet or living specimens may be examined *directly* without desiccation.
4.   Such specimens are so transparent, and x-rays have such penetrating power that thicknesses of 20–50 μm are permissible.
5.   If parts of a specimen differ (or are treated to differ) in atomic number, there is plenty of contrast.

These microscopical advantages would pertain if x-rays could be focused, but their limited refraction precludes the use of lenses. In addition, the absence of polarity on x-ray photons precludes the use of electrostatic or magnetic fields.

Contact radiographs such as a dentist employs are useful if a magnification of 10–20× is sufficient. Higher magnifications are not useful because the contact image is insufficiently sharp, due to the fact that the source of x-rays (electrons) is too broad a beam.

By focusing the beam of electrons to a point on the metallic emitter of x-rays, the resultant radiograph is sufficiently sharp to warrant a light-microscopical magnification of 100–200×. Indeed, a higher magnification is limited by the grain of the photograph rather than the blur of the image.

Since x-ray micrographs are pictures of absorption, they may give better contrast than light micrographs of inorganic parts or particles of metals, ores, minerals, rocks, ceramics, silicones, pigments, and fillers. Moreover, elements of high atomic number may be added as selective stains to parts of organic systems[8,9] for automatic analysis.

Acoustic Microscopy

16.1. ULTRASOUND WAVES FROM MICROSPECIMENS

Acoustic microscopes, the newest members of the microscopical family, employ electromagnetic signals and transform them into *acoustic* waves by means of a piezoelectric transducer. The resultant acoustic waves are of the same propagating elastic type as water waves in the ocean and sound waves in air.[1] Being of very high or ultrahigh frequency, 100–1000 megacycles/sec or megahertz (MHz), the microscopical acoustic waves are ultrasonic; they cannot be heard. As in macroscopical use of ultrasound,[2] the primary significance is that such acoustic waves are reflected or deflected by the specimen's variations in *density* or *stiffness* rather than by differential refraction, absorption, or reflection (as is the case with light or electrons). Consequently, a whole new microscopical area of structural information is now open for correlation with properties or behavior and composition or treatment.

The specimen is contained in a cell filled with a liquid, often water. Incident acoustic waves are deviated by the specimen and are collected and converted to electronic signals which are translated into an image on a cathode-ray tube, in video fashion.

16.2. ACOUSTIC MICROSCOPES: SLAM VERSUS SAM

At the present time there are two versions of the scanning acoustic microscope, which are classified according to how the scanning is performed. In the Sonomicroscope® scanning is done in conjunction with a laser to make it a scanning *laser* acoustic microscope (SLAM). As shown in Figure 16.1, the specimen is mounted in shallow liquid and is held on a stationary stage which is activated acoustically at 100 or 500 MHz. The specimen is covered with a mirrored semitransparent cover

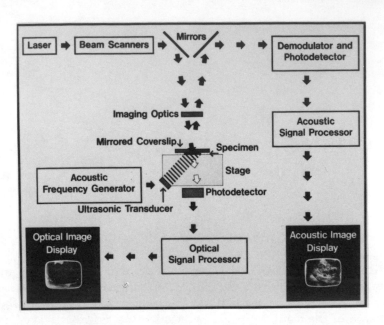

FIGURE 16.1. Simplified block diagram of the Sonomicroscope®, scanning-laser acoustic microscope (SLAM).[3]

slip. The resultant differential disturbances of the mirrored slip are detected by the scanning laser (*flying* spot) and are converted into electrical signals by a photodetector. The (weak) signals are amplified and decoded so as to produce an acoustic image display (lower right of Figure 16.1). Simultaneously, an optical image can be produced with the small amount of laser light that may be transmitted differentially and displayed (lower left of Figure 16.1), or differentially reflected from the specimen and displayed (not specified in Figure 16.1). Hence, a special feature of SLAM is that the laser-light image may be compared simultaneously with the acoustic image.[3]

An optional feature of the Sonomicroscope® (Figure 16.2) is the feeding of acoustic and optical signals into separate channels of a color TV monitor. Thus the separate identities of the acoustic and optical images are maintained in contrasting colors. This feature is important in interpreting the novel acoustic image in terms of the conventional optical image. Interference patterns may be produced by means of an electronic phase reference. The result is a graphic display of changes in density and elasticity within the specimen.[3]

In the scanning acoustic microscope without laser (SAM), scanning is performed *mechanically* by moving the specimen, as shown in Figure 16.3.[4]

Figure 16.4 (left) shows that an electromagnetic input signal is *transformed* into an *acoustic* beam by a piezoelectric film of zinc oxide on the input sapphire crystal. The acoustic beam travels through the liquid (water) in the cell which also contains the object, supported by Mylar® polyester film. The surfaces of the input sapphire crystal and the output sapphire, which are in contact with the spherical part of the specimen's cell, are concave. Thus the input acoustic beam is focused (condensed) on the specimen, which scatters the output acoustic beam into the output (objective) lens (liquid-sapphire) system. The output acoustic beam is converted into the output electromagnetic signal by the respective piezoelectric film transducer (Figure 16.4, right). The acoustic cell at its waist constricts the acoustic beam to less than one wavelength (approximately 2 μm) in diameter. The object is scanned in a raster pattern. The acoustic output modulates the intensity of the output electrical signal and displays it on a cathode-ray tube (CRT, Figure 16.3). The resultant image can correspond either to amplitude modulation or phase interference. In Figure 16.5a, the amplitude (intensity) mode is shown in an acoustic micrograph of a fixed preparation of red blood cells. The change in amplitude of the acoustic beam gives information regarding acoustic absorption and impedence. On the other hand, in Figure 16.5b (the phase mode), there is a point-by-point display of changes in the velocity of sound when the section is of uniform thickness.[1,4–7]

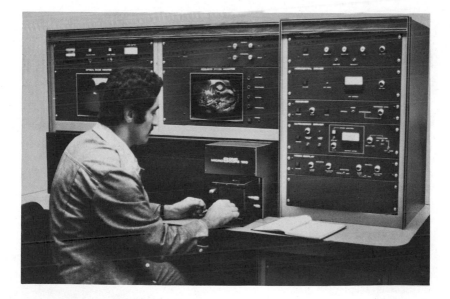

FIGURE 16.2. Photograph of the SLAM (scanning-laser acoustic microscope) Sonomicroscope® for producing acoustic micrographs of specimens.[3]

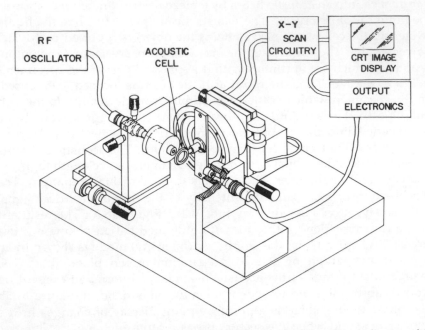

FIGURE 16.3. View of the scanning acoustic microscope (SAM) with an indication of the scanning drives and associated electronic equipment.[4,5]

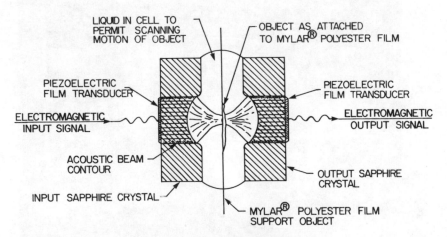

FIGURE 16.4. Schematic representation of the acoustic cell as used in the scanning acoustic microscope (SAM).[1,4]

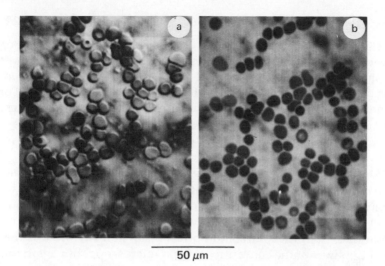

50 μm

FIGURE 16.5. Scanning acoustic micrographs (a) amplitude modulation, (b) phase interference, of red blood cells fixed with methanol. 1000 MHz. May 28, 1976. Courtesy of C. F.
Quate.[7]

The SAM is being applied to medical problems such as malignance
in blood cells, mammillary glands, lymph glands, and lung tissue.[4]
Problems involving even greater constituent differences in density and
stiffness could include teeth, bones, and various types of kidney stones.
Foreign microscopic particles involved in diseases of the lungs include
siliceous minerals, asbestoses, other fibers, soil minerals, and air pollut-
ants. The physiologically detrimental compounds of lead and mercury
also seem to be acoustically determinative.

In nonbiological fields, acoustical studies are being made on solid-
state materials such as a thin aluminized pattern deposited on silicon.
Another example is a circuit of silicon on sapphire, in which microscopic flaws were discovered solely by acoustic microscopy. In cold-
worked aluminum, acoustic micrographs revealed not only the surfacial
slip lines but also the structural changes below the surface.[4] Other
variations in stiffness and density may be revealed among constituents
in alloys, welds, ores, slags, ceramics, cements, pigmented materials,
and reinforced plastics, among others.

Conceivably, the acoustic frequency in the SLAM could be changed
to improve the resolving power. Still, the resolving power of the SLAM,
even at 500 MHz, depends on the distance between the object plane of
interest and the cover slip, since the SLAM operates in a manner analogous to x-microradiography.

16.2.1. Theoretical Resolving Power of the Acoustic Microscope

The theoretical *resolving power* of the acoustic microscope of course depends on the acoustical wavelength in a particular medium. Table 16.1 shows how the wavelength at 100 MHz varies with the medium, depending on its density and stiffness. The resolving power of the Sonomicroscope® 100 (MHz) type of SLAM is stated to be between 0.7 and 1.3 wavelengths.[3] In the Sonomicroscope® model 500, the input frequency is 500 MHz with a correspondingly higher theoretical resolving power.[3]

At present, a still higher frequency in the SAM (1000 MHz) ensures a correspondingly higher theoretical resolving power. In water the wavelength is 1.5 μm and the theoretical resolving power is possibly half of that. The net optical resolvability is less than the audio, but comparable. Both the audio and optical resolvabilities are less than the resolving power of the conventional light microscope, but again they are comparable.[1] Moreover, there is promise of increasing the operating frequency to 2000 MHz, thereby resolving points that are only 0.25 μm apart. There are also plans to employ liquid helium instead of water so as to reduce the acoustic velocity to as little as one-eighth and to yield a resolution near 100 nm (1000 Å).[1]

16.2.2. Practical Resolution of the Acoustic Microscope

In addition to the wavelength, the practical *resolution* by means of an acoustic microscope depends primarily on the effective diameter of the detector. In the lensless SLAM the laser beam diameter may be

TABLE 16.1
Typical Wavelengths at 100 MHz, in Micrometers (μm)[3]

Material	Compressional waves	Shear waves
Air	4	
Water	15	
Soft tissue	14–17	
Hard tissue	30–60	
Plastic–acrylic	27	11
Metals		
Aluminum	63	31
Brass	43	20
Iron	59	32
Titanium	60	31
Glass	56	34

adjusted to the limit of the optical wavelength. In the laserless SAM the diameter of the acoustic beam at the waist of the cell (Figure 16.4) is stated to be less than one wavelength in water, i.e., <1.5 μm.[1] The resulting limit of resolution is about 2 μm. Water is used as mounting liquid, not only because it is compatible with biological and other specimen types, but also because it supports a sufficiently low acoustic velocity to effect good practical resolution at a given input frequency. At the same time, water has high enough absorption to prevent stray multiple acoustic reflections from degrading the image (provided that the input and output electrical connections are well shielded from one another).[4]

For maximum practical resolution, all relevant adjustments must be at their best. In the SAM the two lenses (condenser and objective) must be carefully aligned in all three directions, since the beam's diameter is so small ($\approx$1 μm).

In both types of acoustic microscopes practical resolution also depends upon maintenance of the frequency of incident ultrasound. In the SLAM the laser beam scanner must be sharply controlled, and in laserless SAM the mechanical scanning mechanism must be functioning within certain specifications.

16.2.3. Contrast in Acoustic Images

Contrast is generally adequate in acoustic images because the variations in elastic parameters are greater than those in optical parameters. Furthermore, in either the SLAM or the SAM electronic amplification of contrast is readily available within the bounds of the signal-to-noise ratio.[1] Both types of acoustic microscopes provide an interference mode of operation by comparing either the transmitted or the reflected signal with a constant portion of the initial electrical oscillation.[3,5] The resulting phase interference micrograph may then be compared with the amplitude (intensity) micrograph as a study in contrasts. The use of color contrast between acoustic and laser micrographs has already been mentioned as an optional feature of the SLAM.[3]

16.2.4. Aplanatic Lenses in SAM

Aberrations in the spherical lens of the cell (Figure 16.4) are negligible at this stage in development of the mechanical SAM. In fact, the lens is inherently *aplanatic*, because the concave spherical face of the sapphire contacts water instead of air. The acoustic velocities of water and sapphire are in the ratio of 7.45:1, so that the acoustic rays are sharply refracted at an angle closely parallel to the radii. The focal length of the lens is only 15% greater than the radius[6] (very low f number; very high NA).

16.2.5. Cleanliness in Acoustic Microscopes

Cleanliness must be a factor in both types of acoustic microscopes because the specimen is immersed in the liquid medium (e.g., water). Solutes bleeding from the specimen into the liquid medium might change the velocity of sound significantly, and local concentrations of solute would introduce aberrations. Such considerations pertain largely to biological specimens which might contain a natural serum, an isotonic solution, or a "fixing" reagent, unless it were cleaned of such matter. Of course, when light is introduced into either type of acoustic microscope, any turbidity becomes an optical problem.

16.2.6. Depth of Focus in SAM

The *depth of field in focus* (depth of penetration) in the SAM is probably on the order of 1–5 μm. However, with a "thick" section (6–8 μm) the acoustical image is not quite as sharp as that of a thin (1 μm) section, although more information is obtained about differential densities and stiffness. But with opaque or relatively dense materials such as metals, inorganic glass, kidney stones, or bone, the image comes chiefly from the surface and therefore is in sharp focus.[6] The SLAM accommodates samples of up to several millimeters thick. The acoustic penetration is generally inversely proportional to the acoustic frequency of the specimen (Table 16.1). Evidently the depth of field in focus is greater in the SLAM (at 100 or 500 MHz) than in the SAM (at 1000 MHz).

16.2.7. Focusing SAM

Focusing the SAM is done by a micrometer screw holding the specimen in the cell (Figure 16.3). When the object plane coincides with the narrowest cross section of the acoustic beam, the sharpest focus is achieved.[4]

16.2.8. Acoustic Radiation

The acoustic *radiation* which is put into the Sonomicroscope® 100 is at a frequency of 100 MHz (or, as an option, 500 MHz).[3] As shown in Table 16.1, compressional (and in some materials, shear) waves of characteristic wavelength are set up in the mounting medium and in the specimen. Coincidentally, a flying spot of laser coherent light scans the microfield, as shown in Figure 16.1. The flying spot detects the perturbations of the mirrored slip covering the vibrating specimen. A photodetector converts the signaling laser beam into electric signals which are amplified and decoded into the amplitude- (intensity-) modulated acoustic micrograph. At the same time, some of the incident light of the laser beam is reflected and/or transmitted by the specimen. Either the reflected or transmitted light beam may be photodetected and displayed by a separate video system. The resultant optical micrograph may be compared with or superimposed upon the SLAM acoustic micrograph.[3]

The *radiation* currently put into the SAM is 1000 MHz, and experiments are being made with frequencies of up to 2000 MHz in attempts to increase the resolving power.[1] The *solid angle of incidence* for the acoustic condenser lens corresponds to an *f*-number of 0.65.[4] The acoustic cell (Figure 16.4) may be tilted 90° to place the acoustic axis in a vertical position. Thus the entire cell may be immersed in a medium suitable for maintaining the cell's environment in a viable condition. Then living cultures may be viewed over an extended period of time.[2] The 90° tilt places the specimen in a horizontal position. With the condenser (input lens) on top, the acoustic microscope is like an inverted light microscope, with some of the same advantages.

The *incident* angle in the Sonomicroscope® is normally 10° for aqueous media. Stages with other incident (insonification) angles are optional.[3] In the SAM the incidence is axial (0°).

16.2.9. Magnification

The *magnification* quoted[8] for the Sonomicroscope® "100" is 100×; for the "500" it is quoted as 5000×. At a practical resolution of 20 μm, a magnification of 10 or 20× would be sufficient to meet the minimum requirement of the eye, 200 μm. At a resolution of 4 μm for the "500," 50 or 100× would be useful. How much more magnification is useful rather than empty (see Chapter 2) would depend on the projected use (such as in micrometry). The quoted[8] magnification of the SAM is 500–1000×. At a practical resolution of 2 μm, 100 or 200× would be sufficient for the eye. How much more magnification is useful would depend on other uses besides simple qualitative visual examination. Obviously, if the theoretical resolution of 0.5 μm is achieved, 500–1000× is useful magnification.

16.2.10. Field of View

The *field of view* with the Sonomicroscope® SLAM is 1–12 mm²,[1,8] and 30 images are produced each second on the TV monitor, making it possible to observe dynamic activity. The field with the SAM is much smaller, 0.25 mm on a side. This area amounts to 5×10^4 "elements" of information and takes about 1 sec to scan.[4,8] Larger areas are composited (Figure 16.6).[6]

16.2.11. Stray Acoustic Radiation

Stray acoustic radiation produced in the SAM by multiple reflections is attenuated enough to avoid degradation of the image ("glare"), if care is taken to shield the electrical *in* and *out* connections from each other.[4]

16.2.12. Three-Dimensional Aspect of SLAM

The *three-dimensional aspect* of the SLAM is heightened by superimposing the acoustical and optical micrographs.[9,10] The

FIGURE 16.6. Human liver tissue as invaded by hepatitis. Acoustic frequency of 900 MHz in the SAM.[6] (Sample prepared and analyzed in the laboratory of Dr. J. Edward Berk.)

Sonomicroscope® SLAM accommodates any sample, transparent or reflecting, as thick as several millimeters.[3]

16.2.13. Thickness of the Specimen

The *depth* of "thick" biological sections for use in the SAM is 6–8 μm, often cut from frozen fresh samples. The depth of "thin" sections, generally cut from mounted fixed specimens, is about 1 μm.[6] Reflective materials such as metals, their compounds, coal, cements, and ceramics are studied at their surfaces. Thickness of the specimen is then a dimension of convenience.

16.2.14. Working Distance

Working distance above a specimen in the SLAM is apparently ample for experimenting with the specimen *in situ*. The specimens may be placed directly on the stage or on a glass slide; however, they are covered with a semireflective cover slip. Interchangeable stages are available for the Sonomicroscope® SLAM for polarized shear-wave insonification, alternate insonification angles, and other special work.[3] In the SAM, the specimen is mounted on a Mylar® polyester membrane that is stretched across a thin metal ring (Figure 16.3). The ring is connected to the movement of a loudspeaker cone. The ring assembly is enclosed in the cell, which is filled with a liquid (Figure 16 4).[4] The consequence is that there is no practical working distance beyond the specimen.

16.2.15. Structure of the Specimen

Structure contributes fundamentally to acoustic microscopy, since the differentiating features of the image represent microvariations in density, stiffness, viscosity, or viscoelastic properties. Table 16.1 shows certain types of matter that differ greatly in acoustic properties. Even in microscopic amounts such different materials as air, water, soft tissue, hard tissue, plastic, elastomer, metal, or inorganic glass will be sharply contrasted against one another in their composite acoustical image. Indeed, brass is a composite in itself, and the various possible constituents such as α and β solid solutions of copper and zinc are probably distinguishable, since the former is based on the crystallographic space lattice of copper and the latter on that of zinc. Iron can be present as α, γ, or other crystallographic structure. Acoustic absorption probably differs not only with the iron's structure but also with the amount of carbon in solid solution in steels. Moreover, alloy steels vary in crystal structure and physical properties with the kind and amount of alloying element such as vanadium, chromium, or nickel. There is, perhaps, no more dramatic example of the effect of structure on elastic properties than that of carbon crystallized either as diamond or graphite. Other industrial examples of crystalline allotropy are the white pigment TiO_2 as rutile and/or anatase; $CaCO_3$ as calcite, aragonite, or vaterite.

16.2.16. Anisotropy

Anisotropy is a related structural phenomenon that may be acoustically important. Variations in elasticity according to direction of propagation of the ultrasound waves may be due to anisotropy in particular molecules, crystals, glasses, fibers, films, or other solids or semisolids.

Change in structure has been investigated in stretched and unstretched elastin or collagen fibers.[1] The same kind of experiments would be interesting with elastomeric fibers and with strained polymers.

16.2.17. Morphology of the Specimen

The study of *morphology* in both the SAM and the SLAM is limited by their respective resolving powers. Hence, in pioneering research at this stage of development, it may be well to choose systems or problems involving relatively large microscopic morphology.

16.2.18. Information about the Acoustical Image

Relevant *information* is inherent in the SLAM, since laser light images are readily available and superimposable on acoustical images.[3,11] With the SAM, ordinary light images are independently formed.[1,4,5] In either case, information accumulated during the long history of light microscopy (Chapter 1) is made available for the interpretation of acoustical images. Likewise, macroscopical information about elastic properties and strength of materials should be considered when interpreting acoustical images.

16.2.19. Experiments with the Specimen

Among all the dynamic *experiments* which may be performed with acoustic microscopes, the most relative would be studies of the mechanism of hearing in humans, land and aquatic animals, insects, and other interesting creatures. Aquatic studies are especially relevant because of the use of water as the mounting medium. Indeed, biological material in general may be studied conveniently *in vivo* or *in vitro*.[1,6] Inanimate aqueous or hydrophylic material is likewise amenable to acoustical experimentation. Examples may be wet-strengthened paper, wet fibers, tow, yarn, cloth, wood, or sponge. It is conceivable that such specimens may be dynamically stressed *in situ*. Acoustic studies of the stiffening of fibers (dyed as well as undyed), statically and dynamically at various temperatures, would seem to be logical experiments. Similar studies with other plastic materials, especially those containing pigments, carbon black, pulverized minerals, or man-made fibers, are needed by industry. Hydrophobic materials might be studied in a nonaqueous liquid suitable in the SAM cell. Metallic fibers, powders, foils, wires,[12] and fabrics could be studied acoustically with regard to cold working, annealing, plating, welding, alloying, and compositing. Copper-plated paper laminates, widely used as microcircuits in the electronic industries, present many problems that might be solved uniquely with the SAM. Solid-state materials and devices already have been examined.[5,13]

16.2.20. Behavior of the Specimen

Behavior of disease on tissue has been studied in the SAM by post-mortem examination and by sampling live objects.[6] Comparison of the elasticity and density of diseased tissue with that of healthy tissue is a capability that is unique with acoustic microscopes. Also unique is the opportunity to compare changes in elasticity and density (by audio scanning) with changes in optical parameters (by coherent[3] or non-coherent[1,4,5] light). In the SLAM the two kinds of images may be compared by separate displays or by a single superimposition with differentiating colors, as in locating the advancing front of the disease.[3]

While the study of elasticity and density in most biomedical specimens is new,[14] the study of such properties in industrial materials, even in microscopical specimens, is an old part of physical testing of stress–strain relationships. In the acoustic microscope, however, we have a new and distinctive feature which allows us to view directly the changes in elasticity. For example, it is well known that natural rubber ages poorly, particularly in oxygen or ozone; with an acoustic microscope the actual aging may be viewed. Changes in elasticity may be studied microscopically in other materials such as plastics, metals, alloys, ceramics, and inorganic glasses.[13]

16.2.21. Preparation of the Specimen

The *preparation* for SLAM or SAM of a polished thick section of a specimen such as metal, ore, rock, plastic, bone, or teeth is similar to the preparation for examination by reflected light (see Chapter 4). The only special requirement is that the so-called "thick" section be only a few millimeters thick.

For thin (1-μm) microtomed sections, the less preliminary the preparation such as dehydration and fixation, the better, until it is learned what effect the treatment has on the specimen's elastic properties. Microtomy of a frozen specimen, however, was found to be more difficult than fixing, mounting in, and releasing from poly(methyl methacrylate).[6]

16.2.22. Photomicrography

Photomicrography is presently provided by the Polaroid® type of camera back and automatic fast-developing photographic film.[3] With the advantages of speed, simplicity, and convenience of the Polaroid® system come the restrictions on choice of photosensitive materials, variations in photoprocessing, multiple photoprinting, and transparencies for projection.

Photomicrography by means of the video image[4-6] brings limitations to resolution caused by the raster lines. A video tape recorder[3] may relieve some of that limitation at least during storage of the information, and at the same time permit dynamic events to be recorded such as cardiovascular contraction[14] and the fracture process in steel.[12]

16.3. SUMMARY

At present, there are two kinds of acoustic (ultrasound) microscopes. One kind employs a scanning laser beam microphone and is therefore called a scanning laser acoustic microscope, or SLAM.[3] The other acoustic microscope employs mechanical movements[4,5] so that its code letters are simply SAM.

The SLAM is lensless; it is an acoustic shadowgraph projection microscope whose resolving power is limited by the wavelength of the ultrasound and the diameter of the laser flying spot. With an input frequency of 100 MHz the limiting resolution of the SLAM is about 20 μm; with an input of 500 MHz the limiting resolution is about 4 μm.[3]

The SAM possesses two small identical planoconcave lenses: One acts as a condenser of the acoustic beam, and the other acts as the objective. Both lenses are of sapphire crystal, immersed in a liquid, usually water (Figure 16.1). Although each lens is of a single component, it is practically aplanatic, because the ratio of acoustic velocity in sapphire to that in water is so high (7.45 : 1). As a result, the acoustic rays are very sharply focused. Since the focal length of each water-immersion lens is only 15% longer than the radius, the numerical aperture is very high. The velocity of sound in water is so low that the wavelength (λ) at a frequency of 1000 MHz is only 1.5 μm. This corresponds to a limiting resolving power in the SAM of 0.5 μm.[6]

After characteristic interaction with the specimen in the SAM, acoustic energy is converted by a piezoelectric transducer to an electromagnetic signal which is imaged on a TV screen. In the SLAM the output acoustical beam modulates the laser beam as it scans the field. The modulated laser beam is received by a photodetector that yields two signals: One depends on the optical features of the specimen; the other is proportional to the *acoustic* amplitude from within the specimen. Each type of signal is TV-monitored separately, but the two images may be superimposed, even in contrasting color. With the laserless SAM, ordinary light is used separately to make an optical image for comparison with the acoustical image of the same field. In both acoustic microscopes the optical image is helpful in the interpretation of the acoustical image, especially during this period of experimental applications.

Acoustic interference microscopy can be incorporated into either instrument by combining the signals from the attenuated acoustic rays with a reference signal. The resultant interferogram consists of light and dark bands. By analyzing the position of the bands, quantitative data on density and elastic properties are derived on a microscale.

Applications of acoustic microscopy in the micron range stem from experience in the macroscopical (millimeter) range with the familiar pulse-echo devices used in medical diagnoses and in material flaw detection. The micron range of resolution is also that of light microscopy (Chapters 1–11). Thus the exploration of *elastic* or *dense* microstructure by ultrasound waves is where light microscopy was four centuries ago. It is now hoped that acoustic microscopy will progress to the areas where electron microscopy began four decades ago.

References

References for the Preface

1. E. M. Slayter, *Optical Methods in Biology*, Interscience Div., John Wiley and Sons, Inc., New York, N. Y. 10016 (1970).
2. S. H. Gage, *The Microscope*, 17th ed., Comstock Publishing Associates, Ithaca, N. Y. 14850 (1947).
3. E. M. Chamot and C. W. Mason, *Handbook of Chemical Microscopy*, John Wiley and Sons, Inc., New York, N. Y. 10016. *Vol. 1: Physical Methods*, 3rd ed. (1958); *Vol. 2: Chemical Methods*, 2nd ed. (1940).
4. G. G. Cocks, Associate Professor, Chemical Engineering, Olin Hall, Cornell University, Ithaca, N. Y. 14853.
5. R. Vallery-Radot, *The Life of Pasteur* (translated from French by R. L. Devonshire), Doubleday, Page and Co., Garden City, N. Y. 11530 (1926).
6. D. W. Humphries, The contributions of Henry Clifton Sorby to microscopy, *The Microscope and Crystal Front* **15**, 351–362 (1967).

References for Chapter 1

1. C. Jacker, *Window on the Unknown, a History of the Microscope*, Charles Scribner's Sons, New York, N. Y. 10017 (1966).
2. S. H. Gage, Brief history of lenses and microscopes, in *The Microscope*, 17th ed., Comstock Publishing Associates, Ithaca, N. Y. 14850 (1947).
3. *Milestones in Optical History* (booklet of paintings), Bausch and Lomb Optical Co. Rochester, N. Y. 14602 (1954).
4. I. Asimov, *Biographical Encyclopedia of Science and Technology*, Doubleday and Co., Garden City, N. Y. 11533 (1972).
5. L. van Puyvelde, *Les Primitifs Flamands*, Editions Marion, Bruxelles, Belgium (1947). See plates 12 and 13, and introductory notes, p. 18, indicating that eyeglasses were in use in 1530.
6. Origin and development of the microscope, in *Catalogue of the Royal Microscopical Society* (A. H. Disney, ed.), The Royal Microscopical Society, Oxford OX4 1AJ, England (1928).
7. M. Rooseboom, The history of the microscope, in *Proceedings of the Royal Microscopical Society* **2**, 266–293 (1967).
8. *Compilation of ASTM Standard Definitions*, 3rd. ed., American Society for Testing and Materials, Philadelphia, Pa. 19103 (1976).

9. M. C. Leikind, the microscope collection of the Medical Museum, Armed Forces Institute of Pathology, in *What's New*, house magazine of the Abbott Laboratories, pp. 17–22, Chicago, Ill. 60064 (December 1952).

10. *The Billings Microscope Collection of the Medical Museum* 2nd ed. (H. R. Purtle and J. A. Ey, Jr., eds.), Armed Forces Institute of Pathology, Washington, D. C. 20305 (1974).

11. R. Hooke, *Micrographia*, Royal Society of England (1665); reprinted by Dover Publications, Inc., New York, N. Y. 10014 (1961).

12. E. M. Chamot and C. W. Mason, *Handbook of Chemical Microscopy, Vol. 1: Physical Methods*, John Wiley and Sons, New York, N. Y. 10016 (1st ed., 1930; 3rd ed., 1958).

13. G. S. M. Cowan, My ivory tower, *The Microscope and Crystal Front* **15**, 132–135 (1966).

14. O. W. Richards, American microscope makers and introduction to the collection, *Journal of the Royal Microscopical Society* **83**, Parts 1 and 2, 123–126 (June 1964).

15. O. W. Richards, Microscopy: yesterday and tomorrow, *Transactions of the American Microscopical Society* **91**, 529–532 (1972).

16. S. Bradbury, The quality of the image produced by the compound microscope: 1700–1840, a monograph on the history of microscopy, in *Proceedings of the Royal Microscopical Society* **2**, 151–174 (1967).

17. S. Bradbury, Now it can be proved, *The Vickers Magazine*, pp. 13–14, Vickers Instruments, Malden, Mass. 02148 (Autumn 1966).

18. E. M. Chamot and C. W. Mason, Chemical microscopy, II. Its value in the training of chemists, *Journal of Chemical Education* **5**, 258–268 (March 1928).

19. D. B. Weiner, *Raspail, Scientist and Reformer*, Columbia University Press, New York, N. Y. 10025 (1968).

20. D. L. Padgitt, *A Short History of the Early American Microscopes*, Microscope Publications Ltd., Chicago, Ill. 60616 (1975).

21. *Encyclopaedia Britannica*, Encyclopaedia Britannica, Inc., Chicago, Ill. 60611 (1973).

22. *Index to Annual Book of ASTM Standards*, American Society for Testing and Materials, Philadelphia, Pa. 19103.

23. British Standard Specification on *Components of Microscopes, Part 2: Dimensions of Microscopes*, British Standards Institution, London, W1A 2BS, England (1963).

24. *ASTM Yearbook*, issued annually by the American Society for Testing and Materials, Philadelphia, Pa. 19103.

25. F. Ruch, Physical techniques in biological research, in *Cells and Tissues*, Vol. 3 (G. Oster and A. W. Pollister, eds.), pp. 149–174, Academic Press, New York, N. Y. 10003.

26. R. Vallery-Radot, *The Life of Pasteur* (translated from French by R. L. Devonshire), Doubleday, Page and Co., Garden City, N. Y. 11530 (1926).

27. D. W. Humphries, The contributions of Henry Clifton Sorby to microscopy, *The Microscope and Crystal Front* **15**, 351–362 (1967).

28. A. H. Bennett, H. Jupnik, H. Osterberg, and O. W. Richards, *Phase Microscopy, Principles and Applications*, Chapter 1, pp. 1–3, for historical background, John Wiley and Sons, New York, N. Y. 10016 (1951).

29. W. Krug, J. Rienitz, and G. Schulz, *Contributions to Interference Microscopy* (translated from German by J. H. Dickson), Hilger and Watts, Ltd., London, N. W. 1 (1964).

30. ASTM Symposium on Interference Microscopy (Abstracts) (P. H. Bartels, ed.), *Materials Research and Standards* **2**, 672 (1962).

31. C. E. Hall, *Introduction to Electron Microscopy*, 2nd ed., McGraw–Hill Book Co., Hightstown, N. J. 08520 (1966).

32. T. Mulvey, The history of the electron microscope, in *Proceedings of the Royal Microscopical Society* **2**, 201–227 (1967).

33. E. F. Burton and W. H. Kohl, *The Electron Microscope, an Introduction to Its Fundamental Principles and Applications*, 2nd ed., Reinhold Publishing Co., New York, N. Y. 10022 (1946).

34. J. H. Reisner, The electron microscope—A symbol of modern science, *Scientific Instruments*, Radio Corporation of America, Camden, N. J. 08100, pp. 14–21 (February 1961).
35. C. J. Burton, R. B. Barnes, and T. G. Rochow, The electron microscope, *Industrial and Engineering Chemistry* **34**, 1429–1439 (1942).
36. T. G. Rochow and R. L. Gilbert, Resinography, in *Protective and Decorative Coatings*, Vol. 5 (J. J. Mattiello, ed.), Chapter 5, John Wiley and Sons, Inc., New York, N. Y. 10016 (1946).
37. M. W. Ladd, article on Electron Microscopy, in *Encyclopedia of Industrial Chemical Analysis*, Vol. 1, pp. 649–685, John Wiley and Sons, New York, N. Y. 10016 (1966).
38. M. W. Ladd, *The Electron Microscope Handbook*, Ladd Research Industries, Inc., Burlington, Vt. 05401 (1973).
39. V. E. Cosslett, Beyond optical limits, *The Vickers Magazine*, pp. 15–16 (Autumn 1966).
40. Electron Microscope Society of America, *Yearbook* (1943), G. G. Cocks, Executive Secretary, School of Chemical Engineering, Cornell University, Ithaca, N. Y. 14850.
41. R. B. Barnes, C. J. Burton, and R. G. Scott, Electron microscopical replica techniques for the study of organic surfaces, *Journal of Applied Physics* **16**, 730–739 (1945).
42. E. Ruska, Past and present attempts to attain the resolution limit of the transmission microscope, in *Advances in Optical and Electron Microscopy*, Vol. I (R. Barer and V. E. Cosslett, eds.), pp. 116–179, Academic Press, New York, N. Y. 10003 (1966).
43. A. Septier, The struggle to overcome spherical aberration in electron optics, in *Advances in Optical and Electron Microscopy*, Vol. 1 (R. Barer and V. E. Cosslett, eds.), pp. 204–227, Academic Press, New York, N. Y. 10003 (1966).
44. J. I. Goldstein and H. Yakowitz (eds.), *Practical Scanning Electron Microscopy*, Plenum Press, New York, N. Y. 10011 (1975).
45. J. W. S. Hearle, J. T. Sparrow, and P. M. Cross, *The Use of the Scanning Electron Microscope*, Pergamon Press, Elmsford, N. Y. 10523 (1972).
46. Sales literature of Sonoscan®, Inc., 752 Foster Ave., Bensenville, Ill. 60106 (1975).

References for Chapter 2

1. *Compilation of ASTM Standard Definitions*, 3rd ed., American Society for Testing and Materials, Philadelphia, Pa. 19103 (1976).
2. T. G. Rochow and E. G. Rochow, *Resinography*, Chapter 1, Plenum Press, New York, N. Y. 10011 (1976).
3. ASTM designation E375-75, *Standard Definitions of Terms Relating to Resinography*, American Society for Testing and Materials, Philadelphia, Pa. 19103.
4. James Thurber in his essay on University Days, in *Thurber's Carnival*, Harper and Row Publishers, Inc., New York, N. Y. 10022 (1945).
5. B. E. Stevenson, *Home Book of Quotations*, 9th ed., Dodd, Mead and Co., New York, N. Y. 10016 (1964).
6. V. Ronchi, Twenty embarrassing questions (translated from Italian by E. Rosen), *Atti della Fondazione di Giorgio Ronchi*, **13**, 3–20, Istituto Nazionale di Ottica (publisher), Series II, No. 822, Florence, Italy (1958).
7. G. Wald, Eye and camera, *Scientific American* **183**, No. 2, 32–41 (August 1950).
8. *Image, Object and Illusion*, a Scientific American monograph published by W. H. Freeman and Company, San Francisco, Calif. 94104 (1974).
9. H. W. Zieler, *The Optical Performance of the Light Microscope*, Part 1, Microscope Publications Ltd., Chicago, Ill. 60616 (1972).
10. E. M. Chamot and C. W. Mason, *Handbook of Chemical Microscopy*, 3rd ed., Vol. 1, John Wiley and Sons, Inc., New York, N. Y. 10016 (1958).

11. ASTM designation E-211, Cover glasses and glass slides for use in microscopy, annual *Index to ASTM Standards*, American Society for Testing and Materials, Philadelphia, Pa. 19103.

12. G. E. Schlueter, Manipulation of depth of field in microscopy, *American Laboratory*, pp. 65, 66, 68, 70 (April 1975).

13. T. G. Rochow, F. G. Rowe, E. J. Thomas, and A. F. Kirkpatrick, Light and electron microscopical studies of pleurosigma angulatum for resolution of detail and quality of image, *The Microscope and Crystal Front* **15**, 177–201 (1966).

14. T. G. Rochow and R. L. Gilbert, Chapter 5 on resinography, in *Protective and Decorative Coatings* (J. J. Mattiello, ed.), John Wiley and Sons, Inc., New York, N. Y. 10016 (1946).

15. R. B. McLaughlin, *Accessories for the Light Microscope*, Microscope Publications, Ltd., Chicago, Ill. 60616 (1975).

16. H. W. Zieler, *The Optical Performance of the Light Microscope*, Part 2, Microscope Publications, Inc., Chicago, Ill. 60616 (1973).

17. W. C. McCrone, *Dispersion Staining*, Microscope Publications, Ltd., Chicago, Ill. 60616 (in preparation).

18. S. Leber, Light source in microscopy, *American Laboratory*, pp. 49, 50, 52 (April 1976).

19. N. H. Hartshorne, *The Microscopy of Liquid Crystals*, Microscope Publications Ltd., Chicago, Ill. 60616 (1974).

20. N. H. Hartshorne and A. Stuart, *Crystals and the Polarising Microscope*, 4th ed., Edward Arnold, Ltd., London, W.1. (1970).

21. L. Wilson, Photomicrographs of rods and cones of the human eye, LIFE magazine, pp. 58–59 (October 1, 1971).

22. G. E. Schlueter and W. E. Gumpertz, The stereomicroscope: Instrumentation and techniques, *American Laboratory*, pp. 61–64, 66 (April 1976).

23. ASTM designation E-210, Microscope objective thread, annual *Index to ASTM Standards*, American Society for Testing and Materials, Philadelphia, Pa. 19103.

24. H. Osterberg and L. W. Smith, Defocusing images to increase resolution, *Science* **134**, 1193–1196 (1961).

25. R. P. Loveland, *Photomicrography*, Vols. 1 and 2, John Wiley and Sons, Inc., New York, N. Y. 10016 (1970).

References for Chapter 3

1. H. W. Zieler, *The Optical Performance of the Light Microscope*, Part 1, Microscope Publications Ltd., Chicago, Ill. 60616 (1972).

2. E. M. Chamot and C. W. Mason, *Handbook of Chemical Microscopy, Vol. 1: Physical Methods*, 3rd ed., John Wiley and Sons, Inc., New York, N. Y. 10016 (1958).

3. G. E. Schlueter and W. E. Gumpertz, The stereomicroscope: Instrumentation and techniques, *American Laboratory*, pp. 61–64, 66, 68–71 (April 1976).

4. R. B. McLaughlin, *Accessories for the Light Microscope*, Microscope Publications, Ltd., Chicago, Ill. 60616 (1975).

5. R. G. Scott, The structure of synthetic fibers, ASTM Symposium on Microscopy, ASTM Special Technical Publication 257, pp. 121–131, American Society for Testing and Materials, Philadelphia, Pa. 19103 (1959).

6. T. G. Rochow and R. L. Gilbert, Resinography, in *Protective and Decorative Coatings* J. J. Mattiello, ed.), Vol. 5, pp. 536–537; 623, John Wiley and Sons, Inc., New York, N. Y. 10016 (1946).

7. N. Meyers, Microscopy needs better microscopists, *Scientific Research*, p. 37 (November 25, 1968).

8. W. C. McCrone, Whither microscopy?, *American Laboratory*, pp. 11–14 (April 1975).
9. S. Bradbury, revision of Peacock's *Elementary Microtechnique*, 4th ed., Edward Arnold, Ltd., distributed by Crane, Russak and Co., New York, N. Y. (1973).
10. S. H. Gage, *The Microscope*, 17th ed., Comstock Publishing Associates, Ithaca, N. Y. 14850 (1947).
11. T. G. Rochow, a review of the International Conference on the Role of the Microscope in Scientific Investigation, Imperial College, London, July 18–22, 1966, *Applied Optics* **6,** 238, 256, 266 (1967).
12. ASTM designation E211, Specifications and methods of test for cover glasses for use in microscopy, *Index to Annual Book of ASTM Standards*, American Society for Testing and Materials, Philadelphia, Pa. 19103.
13. T. G. Rochow, F. G. Rowe, E. J. Thomas, and S. F. Kirkpatrick, Light and electron microscopical studies of *Pleurosigma angulatum*, *The Microscope and Crystal Front* **15,** 177–201 (1966).
14. H. Osterberg, Microscope imagery and interpretations, *J. Optical Society of America* **40,** 295–303 (1950).
15. A. H. Bennett, H. Jupnik, H. Osterberg, and O. W. Richards, *Phase Microscopy*, John Wiley and Sons, Inc., New York, N. Y. 10016 (1951).
16. W. C. McCrone, *Dispersion Staining*, Microscope Publications, Ltd., Chicago, Ill. 60616 (in preparation).
17. H. W. Zieler, The Optical Performance of the Light Microscope, Part II, Microscope Publications, Ltd., Chicago, Ill. 60616 (1973).
18. A. A. Thaer and M. Sernetz, *Fluorescence Techniques in Cell Biology*, Springer-Verlag, New York, N. Y. (1973).
19. C. H. Burch, *Proceedings of the Physical Society of London* **49,** 41ff (1947).
20. R. C. Gore, Infrared spectrometry of small samples with the reflecting microscope, *Science* **110,** 710–711 (1949).
21. J. McArthur, The history of the miniature microscope, in *Proceedings of the Royal Microscopical Society*, **2,** Part 2 (1967).
22. I. M. Roberts and A. M. Hutcheson, Handling and staining epoxy resin sections for light microscopy, *J. Microscopy* **103,** Part 1, 121–126 (January 1976).
23. H. Freund (ed.), *Handbuch der Mikroskopie in der Technik*, Vol. 6, Part 1, Umschau Verlag, Frankfurt/Main, Germany (1972).
24. G. L. Clark (ed.), *The Encyclopedia of Microscopy*, Reinhold Publishing Corp., New York, N. Y. 10001 (1961).
25. C. Stolinski and A. S. Breathnach, *Freeze-Fracture Replication of Biological Tissues*, University of London, England (December 1975/January 1976).
26. Controlled environment culture chamber with air stream incubator, in *Proceedings of the Royal Microscopical Society* **10,** Part 5, 289 (1975).

References for Chapter 4

1. T. G. Rochow and R. L. Gilbert, Chapter 5, Resinography, in *Protective and Decorative Coatings* (J. J. Mattiello, ed.), Vol. 5, John Wiley and Sons, Inc., New York, N. Y. 10016 (1946).
2. *Metallography—A Practical Tool for Correlating the Structure and Property of Materials*, ASTM Special Technical Publication 557, American Society for Testing and Materials, Philadelphia, Pa. 19103 (1975).
3. J. H. Richardson, *Optical Microscopy for the Materials Sciences*, Marcel Dekker, New York, N. Y. 10016 (1971).

4. T. G. Rochow, Microscopy, in *Encyclopedia of Industrial Chemical Analysis*, Vol. 2, pp. 586–590, John Wiley and Sons, Inc., New York, N. Y. 10016 (1966).
5. J. B. Nelson, Apparatus for the microscopical determination of ore minerals, *American Laboratory*, pp. 81–82, 84, 86, 88–91 (April 1975).
6. W. C. McCrone, Surface characterization by light microscopy, preprints of papers presented at the 171st meeting, *ACS, Div. Organic Coatings and Plastics Chemistry* **36**, No. 1, p. 2, New York, N. Y. (April 5–9, 1976).
7. T. G. Rochow and E. G. Rochow, *Resinography*, Plenum Press, New York, N. Y. 10011 (1976).
8. D. W. Skalla and S. R. Mather, A non-destructive method of observing and recording wear characteristics of phonograph records, Inter/MICRO-75, Chicago, Ill., *The Microscope* **23**, 55–60 (1975).
9. E. M. Slayter, *Optical Methods in Biology*, Wiley-Interscience, New York, N. Y. 10016 (1970).
10. V. P. Miniutti, Reflected-light and scanning electron microscopy of ultraviolet irradiated redwood surfaces, *The Microscope* **18**, 61–72 (January 1970).
11. H. W. Zieler, *The Optical Performance of the Light Microscope*, Part 2, Microscope Publications, Ltd., Chicago, Ill. 60616 (1973).
12. F. G. Rowe and H. F. Nicolaysen, Reflected light microscopy of waxes on paper, *International Microscopy Symposium* (W. C. McCrone, ed.), McCrone Associates, Chicago, Ill. 60616 (1960).
13. E. M. Chamot and C. W. Mason, *Handbook of Chemical Microscopy, Vol. 1: Physical Methods*, 3rd ed., John Wiley and Sons, Inc., New York, N. Y. 10016 (1958).
14. R. B. McLaughlin, *Accessories for the Light Microscope*, Microscope Publications, Ltd., Chicago, Ill. 60616 (1975).
15. H. W. Zieler, *The Optical Performance of the Light Microscope*, Part 1, Microscope Publications, Ltd., Chicago, Ill. 60616 (1972).
16. ASTM designation E2, Micrographs of metals and alloys (including recommended practice for photography as applied to metallography), annual *Index to ASTM Standards*, American Society for Testing and Materials, Philadelphia, Pa. 19103.
17. S. Leber, Light source for microscopy, *American Laboratory*, pp. 49, 50, 52 (April 1976).
18. E. N. Cameron, *Ore Microscopy*, John Wiley and Sons, Inc., New York, N. Y. 10016 (1961).
19. ASTM designation E112, Estimating average grain size of metals, annual *Index to ASTM Standards*, American Society for Testing and Materials, Philadelphia, Pa. 19103.
20. ASTM designation E384, Test for microhardness of materials, annual *Index to ASTM Standards*, American Society for Testing and Materials, Philadelphia, Pa. 19103.
21. ASTM National Committee E-4 on Metallography, scope and organization, in annual ASTM *Yearbook*; standards, in annual *Index to ASTM Standards*, American Society for Testing and Materials, Philadelphia, Pa. 19103.
22. L. D. Nichols, M. H. Mohamed, and T. G. Rochow, Some structural and physical properties of yarn made on the integrated composite spinning system, Part 1, *Textile Research Journal* **42**, 338–344 (June 1972).
23. T. G. Rochow and E. G. Rochow, *Microscopical Methods in Resinography*, Microscope Publications, Ltd., Chicago, Ill. 60616 (in preparation).
24. ASTM designation E407, Microetchants for metals and alloys, annual *Index to ASTM Standards*, American Society for Testing and Materials, Philadelphia, Pa. 19103.

References for Chapter 5

1. *Compilation of ASTM Standard Definitions*, 3rd ed., American Society for Testing and Materials, Philadelphia, Pa. 19103 (1976 and later).

2. E. M. Chamot and C. W. Mason, *Handbook of Chemical Microscopy*, 3rd ed., Vol. 1, John Wiley and Sons, Inc., New York, N. Y. 10016 (1958).
3. N. H. Hartshorne and A. Stuart, *Crystals and the Polarising Microscope*, 4th ed., American Elsevier, New York, N. Y. 10017 (1970).
4. F. D. Bloss, *An Introduction to the Methods of Optical Crystallography*, Holt, Rinehart and Winston, New York, N. Y. 10017 (1961).
5. J. G. Delly, Microscopy's color key, *Industrial Research*, 44–50 (Oct. 1973).
6. E. M. Slayter, *Optical Methods in Biology*, Interscience Div., John Wiley and Sons, Inc., New York, N. Y. 10016 (1970).
7. N. H. Hartshorne, *The Microscopy of Liquid Crystals*, Microscope Publications, Ltd., Chicago, Ill. 60616 (1974).
8. T. G. Rochow and E. G. Rochow, *Resinography*, Plenum Press, New York, N. Y. 10011 (1976).
9. F. Ruch, Physical techniques in biological research, in *Cells and Tissues* (G. Oster and A. W. Pallister, eds.), Vol. 3, pp. 149–174, Academic Press, New York, N. Y. 10003 (1955).
10. H. S. Bennett, The microscopical investigation of biological materials with polarized light, in McClung's *Handbook of Microscopical Techniques*, 3rd ed. (R. M. Jones, ed.), pp. 591–677, Hafner Press, New York, N. Y. 10022 (1950).
11. ASTM designation E-(number pending), Method of test for determination of birefringence in fibers of circular cross section by a variable compensator technique, annual *Index to ASTM Standards*, American Society for Testing and Materials, Philadelphia, Pa. 19103.
12. R. B. McLaughlin, *Accessories for the Light Microscope*, Microscope Publications Ltd., Chicago, Ill. 60616 (1975).
13. N. A. Crites, H. Grover, and A. R. Hunter, Photoelastic techniques, *Product Engineering* **33**, Part 3, pp. 57–69 (September 3, 1962).
14. R. C. Emmons in A. N. Winchell, *Microscopic Characters of Inorganic Solid Substances*, Chapter 8, John Wiley and Sons, New York, N. Y. 10016 (1931); also The Universal Stage, Geological Society of America, Boulder, Co. 80301 (1943).
15. ASTM designation E-211, Specifications and methods of test for cover glasses and glass slides for use in microscopy, annual *Index to ASTM Standards*, American Society for Testing and Materials, Philadelphia, Pa. 19103.
16. W. C. McCrone, *Fusion Methods in Chemical Microscopy*, Interscience Div., John Wiley and Sons, Inc., New York, N. Y. 10016 (1957).
17. T. G. Rochow and R. J. Bates, A microscopical automated microdynamometer microtension tester, *ASTM Materials Research and Standards* **12**, No. 4, 27–30 (1972).
18. T. G. Rochow and R. L. Gilbert, Resinography, in *Protective and Decorative Coatings* (J. J. Mattiello, ed.), Vol. 5, Chapter 5, John Wiley and Sons, Inc., New York, N. Y., 10016 (1946).
19. Ward's Scientific Co., sales literature, Wards Natural Science Establishment, P. O. Box 1712, Rochester, N. Y. 14603.

References for Chapter 6

1. A. N. J. Heyn, *Fiber Microscopy*, Interscience Div., John Wiley and Sons, Inc., New York, N. Y. 10016 (1954).
2. *Compilation of ASTM Standard Definitions*, 3rd ed., American Society for Testing and Materials, Philadelphia, Pa. 19103 (1976).
3. AATCC Test Method 20-1973, Fibers in textiles: Identification, *AATCC Technical Manual*, American Association of Textile Chemists and Colorists, Research Triangle Park, N. C. 27709 (1973).

4. P. L. Kirk, *Crime Investigation*, Interscience Div., John Wiley and Sons, Inc., New York, N. Y. 10016 (1953).

5. F. D. Bloss, *An Introduction to the Methods of Optical Crystallography*, Holt, Rinehart, and Winston, New York, N. Y. 10017 (1961).

6. E. M. Chamot and C. W. Mason, *Chemical Microscopy*, 3rd ed., Vol. 1, John Wiley and Sons, Inc., New York, N. Y. 10016 (1958).

7. T. G. Rochow and E. G. Rochow, *Resinography*, Plenum Press, New York, N. Y. 10011 (1976).

8. E. Leitz, *Instructions for Tilting Compensator K*, E. Leitz, Inc., Rockleigh, N. J. 07647 (1971).

9. N. H. Hartshorne and A. Stuart, *Crystals and the Polarising Microscope*, 4th ed., Edward Arnold, Ltd., London, W.1. (1970).

10. M. A. Sieminski, The temperature for zero birefringence of Arnel® and other fibers, *Textile Research Journal* **34**, 918–924 (1964).

11. ASTM designation E (number pending), Method of test for determination of birefringence in fibers of circular cross section by a variable compensator technique, annual *Index to ASTM Standards*, American Society for Testing and Materials, Philadelphia, Pa. 19103.

12. M. A. Sieminski, A note on the measurement of birefringence in fibers, *The Microscope* **23**, 35–36 (1975).

13. R. W. Singleton, M. A. Sieminski, and B. S. Sprague, The effect of radial heterogeneity on fiber properties, *Textile Research J.* **31**, 917–925 (1961).

14. C. P. Saylor, Heterodoxy in refractive index measurement, NYMS Dialogues, May 17–19, 1977, New York Microscopical Society, American Museum of Natural History, New York, N. Y. 10024.

15. Cargille Scientific, Inc., Cedar Grove, N. J. 07009, for example.

16. N. H. Hartshorne, *The Microscopy of Liquid Crystals*, Microscope Publications, Ltd., Chicago, Ill. 60616 (1974).

17. R. G. Scott, A few observations concerning the structure of synthetic fibers, *ASTM Symposium on Microscopy* (F. F. Morehead and R. Loveland, eds.), ASTM STP 257, American Society for Testing and Materials, Philadelphia, Pa. 19103 (1959).

18. Federal Trade Commission, *Rules and regulations* under the Textile Fiber Products Identification Act, March 3, 1969, Federal Trade Commission, Washington, D.C. 20580.

19. A. O. Mogensen, Microscopical apparatus and techniques for observing the fiber-forming process, in *Resinographic Methods*, ASTM Special Technical Publication 348, pp. 31–35, American Society for Testing and Materials, Philadelphia, Pa. 19103 (1964).

20. T. G. Rochow, Some microscopical aspects of resinography, *J. Royal Microscopical Society* **87**, 39–45 (1966).

21. T. G. Rochow and R. J. Bates, A microscopical automated microdynamometer microtension tester, *Materials Research and Standards* **12**, 27–30 (1972).

22. S. Glasstone, *Textbook of Physical Chemistry*, 2nd ed., p. 528, D. van Nostrand Co., 250 Fourth Ave., New York, N. Y. 10013.

23. R. O. Sauer, *J. Am. Chem. Soc.* **68**, 954 (1946).

24. R. West and E. G. Rochow, *J. Am. Chem. Soc.* **74**, 2490 (1952).

References for Chapter 7

1. *Compilation of ASTM Standard Definitions*, 3rd ed., American Society for Testing and Materials, Philadelphia, Pa. 19103 (1976).

2. A. N. Winchell, *The Microscopic Characters of Artificial Inorganic Solid Substances or Artificial Minerals*, John Wiley and Sons, New York, N. Y. 10016 (1938).
3. E. M. Chamot and C. W. Mason, *Handbook of Chemical Microscopy, Vol. 1: Physical Methods*, 3rd ed., John Wiley and Sons, New York, N. Y. 10016 (1958).
4. E. M. Chamot and C. W. Mason, *Handbook of Chemical Microscopy, Vol. 2: Chemical Methods*, 2nd ed., John Wiley and Sons, New York, N. Y. 10016 (1940).
5. H. Behrens and P. D. C. Kley, *Organische Microchemische Analyse*, Voss, Leipzig (1922); translated by R. E. Stevens, *Microscopical Identification of Organic Compounds*, Microscope Publications, Ltd., Chicago, Ill. 60616 (1969).
6. A. N. Winchell and H. Winchell, *The Microscopical Characters of Artificial Inorganic Solid Substances: Optical Properties of Artificial Minerals*, 3rd ed., Academic Press, New York, N. Y. 10003 (1964).
7. C. P. Saylor, *J. Physical Chemistry* **32**, 1441–1460 (1928).
8. R. E. Stevens, Microscopical identification of water in crystals, *Analytical Chimica Acta* **60**, 325–334 (1972).
9. C. W. Mason, Chairman of American Chemical Society Committee on Recommended Practice for Microscopical Reports on Crystalline Materials, in A. C. S. Publications, *Industrial and Engineering Chemistry* (Analytical edition) **17**, 603–604 (1945), and *Analytical Chemistry* **20**, 274 (1948).
10. E. M. Chamot and C. W. Mason, Chemical microscopy, Part 1: Crystallization experiments as an introduction to metallography, *Journal of Chemical Education* **5**, 10–24 (1928).
11. C. W. McCrone, *Fusion Methods in Chemical Microscopy*, Interscience Div., John Wiley and Sons, Inc., New York, N. Y. 10016 (1957).
12. P. L. Kirk, *Crime Investigation*, Interscience Div., John Wiley and Sons, Inc., New York, N. Y. 10016 (1953).
13. A. N. Winchell, *Optical Properties of Organic Compounds*, University of Wisconsin Press, Madison, Wis. 53701 (1943).
14. F. D. Bloss, *An Introduction to the Methods of Optical Crystallography*, Holt, Rinehart and Winston, New York, N. Y. 10017 (1961).
15. N. H. Hartshorne and A. Stuart, *Crystals and the Polarising Microscope*, 4th ed., American Elsevier Publishing Co., New York, N. Y. 10017 (1970).
16. E. E. Wahlstrom, *Optical Crystallography*, 4th ed., John Wiley and Sons, New York, N. Y., 10016 (1969).
17. R. C. Emmons, The universal stage, *Geological Society of America Mem.* **8**, 205 (1943); see also Reference 2, Chapter 8.
18. A. F. Kirkpatrick, unpublished recapitulation of data collected by A. N. Winchell.[6,13]
19. W. C. McCrone, R. G. Draftz, and J. G. Delly, *The Particle Analyst*, Ann Arbor Science Publishers, Inc., Ann Arbor, Mich, 48406, a journal for one year (1968); reprinted as a monograph (1969).
20. W. C. McCrone, A new dispersion staining objective, *The Microscope* **23**, 221–226 (1975).
21. T. G. Rochow, article on Microscopy (Chemical), in the *Encyclopedia of Chemical Technology*, Vol. 13, p. 499, John Wiley and Sons, New York, N. Y. 10016 (1967).
22. *Cyanamid Melamine*, Booklet 1C 9055, American Cyanamid Co., Wayne, N. J., 07470 (1959).
23. M. L. Willard, *Chemical Microscopy*, W. B. Keeler Bookstore, State College, Pa. 16802 (1952).
24. N. H. Hartshorne, *The Microscopy of Liquid Crystals*, Microscope Publications, Ltd., Chicago, Ill. 60616 (1974).
25. F. G. Rosevear, The microscopy of the liquid crystalline neat and middle phases of soaps and synthetic detergents, *J. American Oil Chemists Society* **31**, 628–639 (1954).

26. J. Rogers and P. A. Winsor, Changes in the optical sign of the lamellar phase (G) in the Aerosol OT/water system with composition and temperature, *J. Colloid and Interface Science* **30**, 247–257 (1969).
27. T. G. Rochow and C. W. Mason, Breaking emulsions by freezing, *Industrial and Engineering Chemistry* **28**, 1296–1300 (1936).

References for Chapter 8

1. ASTM designation E375-75, Standard definitions of terms relating to resinography, annual *Index to ASTM Standards,* American Society for Testing and Materials, Philadelphia, Pa. 19103 (1976 and later).
2. *Compilation of ASTM Standard Definitions,* 3rd ed., American Society for Testing and Materials, Philadelphia, Pa. 19103 (1976).
3. R. B. McLaughlin, *Accessories for the Light Microscope,* Microscope Publications, Ltd., Chicago, Ill. 60616 (1975).
4. N. H. Hartshorne, *The Microscopy of Liquid Crystals,* Microscope Publications, Ltd., Chicago, Ill. 60616 (1974).
5. W. C. McCrone, *Fusion Methods,* John Wiley and Sons, Inc., New York, N. Y. 10016 (1957).
6. *Cyanamid Melamine,* Booklet 1C 9055, American Cyanamid Co., Wayne, N. J. 07470 (1959).
7. T. G. Rochow and E. G. Rochow, *Resinography,* Plenum Press, New York, N. Y. 10011 (1976).
8. T. G. Rochow and C. W. Mason, Breaking emulsions by freezing, *Industrial and Engineering Chemistry* **28**, 1296–1300 (1936).
9. S. A. Reigel and R. P. Bundy, High-speed cinematography, *Research/Development,* pp. 24, 25, 26–28 (December 1975).
10. J. Schwartz, Some uses of time-lapse cinemicrography in contemporary research, *American Laboratory,* pp. 37, 39, 40 (April 1970).
11. T. G. Rochow and R. L. Gilbert, Chapter 5, Resinography, in *Protective and Decorative Coatings* (J. J. Mattiello, ed.), Vol. 5, John Wiley and Sons, Inc., New York, N. Y. 10016 (1946).
12. H. W. Zieler, *The Optical Performance of the Light Microscope,* Part 2, Microscope Publications, Ltd., Chicago, Ill. 60616 (1973).
13. ASTM designation E-210, Microscope objective thread, annual *Index to ASTM Standards,* American Society for Testing and Materials, Philadelphia, Pa. 19103.
14. H. W. Zieler, *The Optical Performance of the Light Microscope,* Part 1, Microscope Publications, Ltd., Chicago, Ill. 60616 (1972).
15. D. W. Quackenbush, Critical 35 mm Photomicroscopy, *The Microscope* **23**, 195–211 (1975).
16. R. P. Loveland, *Photomicrography,* Vols. 1 and 2, John Wiley and Sons, New York, N. Y. 10016 (1970).
17. A. Kramer, Direct Steps to Developing, *Modern Photography,* pp. 140–143 (October 1974); Direct Steps to Printing, *Modern Photography,* pp. 144, 190 (October 1974).
18. *Kodak Master Darkroom Guide,* Eastman Kodak Co., Consumer Markets Div., Rochester, N. Y. 14650.
19. E. M. Chamot and C. W. Mason, *Chemical Microscopy,* Vol. 1, 3rd ed., John Wiley and Sons, Inc., New York, N. Y. 10016 (1958).
20. Eastman Kodak Co., Photo Information, Department 841, Rochester, N. Y. 14650.

21. Such as 102 photometer sold by Science and Mechanics, Instruments Division, 229 Park Avenue South, New York, N. Y. 10003, as a kit or assembled.

References for Chapter 9

1. *Compilation of ASTM Standard Definitions,* 3rd ed., American Society for Testing and Materials, Philadelphia, Pa. 19103 (1976).
2. E. G. Rochow, G. Fleck, and T. R. Blackburn, *Chemistry, Molecules that Matter,* Appendix III, Holt, Rinehart and Winston, Inc., New York, N. Y. 10017 (1974).
3. A. F. Kirkpatrick, F. G. Rowe, E. J. Thomas, and T. G. Rochow, Light and microscopical studies of *Pleurosigma angulatum* for resolution of detail and quality of image, *The Microscope and Crystal Front* **15,** 176–201 (1966).
4. L. A. Wren and J. D. Corrington, *Understanding and using the phase microscope,* Unitron Instrument Company, 101 Crossways Park West, Woodbury, N. Y. 11797 (1963).
5. E. M. Chamot and C. W. Mason, *Handbook of Chemical Microscopy,* Vol. 1, 3rd ed., John Wiley and Sons, Inc. New York, N. Y. 10016 (1958).
6. N. H. Hartshorne and A. Stuart, *Crystals and the Polarising Microscope,* 4th ed., American Elsevier, New York, N. Y. 10017 (1970).
7. A. H. Bennett, H. Jupnik, H. Osterberg, and O. W. Richards, *Phase Microscopy,* John Wiley and Sons, Inc., New York, N. Y. 10016 (1951).
8. C. P. Saylor, Chapter 1 in *Advances in Optical and Electron Microscopy* (R. Barer and V. E. Cosslett, eds.), Academic Press, New York, N. Y. 10003 (1966).
9. S. G. Ellis and W. Hunn, High resolution transmitted light phase microscopy of unmounted specimens, *The Microscope* **23,** No. 3, 127–131 (1975).
10. O. W. Richards, The Polanret™ variable densiphase microscope, *Journal of Microscopy* **98,** Part 1, pp. 67–77 (May 1973).
11. Information on the Polanret™ microscope, American Optical Corp., Scientific Instrument Div., Buffalo, N. Y. 14215.
12. R. Hoffman and L. Gross, Modulation Contrast Microscope, *Applied Optics* **14,** 1169–1176 (1975).
13. R. Hoffman, The modulation microscope—Principles and performance, *Journal of Microscopy* **110,** Part 3, pp. 209–219 (August 1977).
14. W. C. McCrone, A new dispersion staining objective, *The Microscope* **23,** 221–226 (1975).
15. G. C. Crossman, *Stain Technology* **24,** 61 (1949).
16. G. C. Crossman, *American Industrial Hygiene Quarterly* **18,** 341 (1957).
17. N. H. Hartshorne and A. Stuart, *Crystals and the Polarising Microscope,* 4th ed., Edward Arnold, Ltd., London, England (1970).
18. R. P. Cargille, Inc., Cedar Grove, N. J. 07009, special set of liquids for dispersion staining.
19. W. C. McCrone and J. G. Delly, *Particle Atlas,* 2nd ed., Ann Arbor Science Publishers, Ann Arbor, Mich. 48106 (1973).
20. W. C. McCrone, reply to Letter to the editor by D. L. Faulkner and T. G. Rochow, *The Microscope* **20,** 228–230 (1972).
21. W. C. McCrone and R. I. Johnson, *Techniques, Instruments and Accessories for Microanalysts, A User's Manual,* Walter C. McCrone Associates, Chicago, Ill. 60616 (1974).
22. ASTM designation E-210, Microscope objective thread, American Society for Testing and Materials, Philadelphia, Pa. 19103.
23. W. C. McCrone, Determination of n_D, n_F and n_C by dispersion staining, *The Microscope* **23,** 213–220 (1975).
24. J. Dodd and W. C. McCrone, A Schlieren eyepiece, *The Microscope* **23,** 89–92 (1975).

References for Chapter 10

1. E. M. Chamot and C. W. Mason, *Handbook of Chemical Microscopy*, Vol. 1, 3rd ed., John Wiley and Sons, Inc., New York, N. Y. 10016 (1958).
2. W. Lang, Nomarski differential interference contrast, *American Laboratory*, pp. 45–46, 48, 50, 52 (April 1970).
3. S. Tolansky, *Multiple-Beam Interference Microscopy of Metals*, Academic Press, New York, N. Y. 10003 (1970).
4. P. Sullivan and B. Wunderlich, *Interference Microscopy of High Polymers*, Office of Naval Research, Technical Report No. 4, Contract Nonr-401 (44), Task No. NR 051-428, Cornell University, Department of Chemistry, Ithaca, N. Y. 14850 (1963).
5. R. B. McLaughlin, *Accessories for the Light Microscope*, Microscope Publications, Ltd., Chicago, Ill. 60616 (1975).
6. ASTM designation E-210, Microscope objective thread, annual *Index to ASTM Standards*, American Society for Testing and Materials, Philadelphia, Pa. 19103.
7. R. G. Scott, A few applications of the interference microscope to the study of fibrous materials (in German), *Leitz Mitteilungen für Wisseschaft und Technik* 5(5), 132–140, Wetzlar, Germany (März, 1971).
8. N. H. Hartshorne and A. Stuart, *Crystals and the Polarising Microscope*, 4th ed., Edward Arnold, Ltd., London, W.1. (1970).

References for Chapter 11

1. *Compilation of ASTM Standard Definitions*, 3rd ed., American Society for Testing and Materials, Philadelphia, Pa. 19103 (1976).
2. T. G. Rochow and R. L. Gilbert, Resinography, in *Protective and Decorative Coatings* (J. J. Mattiello, ed.), Vol. 5, Chapter 5, John Wiley and Sons, Inc., New York, N. Y. 10016 (1946).
3. R. B. McLaughlin, *Accessories for the Light Microscope*, Microscope Publications, Ltd., Chicago, Ill. 60616 (1975).
4. T. G. Rochow, Some microscopical aspects of resinography, *J. Royal Microscopical Society* **87,** 39–45 (1967).
5. T. G. Rochow and R. J. Bates, A microscopical automated microdynamometer microtension tester, *Materials Research and Standards* **12,** pp. 27–30 (April 1972).
6. J. I. Goldstein and H. Yakowitz (eds.), *Practical Scanning Electron Microscopy*, Plenum Press, New York, N. Y. 10011 (1975).
7. E. M. Chamot and C. W. Mason, *Handbook of Chemical Microscopy*, Vol. 1, 3rd ed., John Wiley and Sons, Inc., New York, N. Y. 10016 (1958).
8. ASTM designation E-210, Microscope objective thread, annual *Index to ASTM Standards*, American Society for Testing and Materials, Philadelphia, Pa. 19103.
9. ASTM designation E-384, Standard method of test for microhardness of materials, annual *Index to ASTM Standards*, American Society for Testing and Materials, Philadelphia, Pa. 19103.
10. W. C. McCrone, *Fusion Methods*, John Wiley and Sons, Inc., New York, N. Y. 10016 (1957).
11. N. H. Hartshorne, *The Microscopy of Liquid Crystals*, Microscope Publications, Inc., Chicago, Ill. 60616 (1974).
12. D. G. Grabar and R. Haessly, Identification of synthetic fibers by micro fusion methods, *Analytical Chemistry* **28,** 1586–1589 (1956).
13. E. M. Chamot and C. W. Mason, Chemical microscopy. I. Crystallization experiments as an introduction to metallography, *J. Chemical Education* **5,** pp. 9–24 (January 1928).

14. F. D. Bloss, *An Introduction to the Methods of Optical Crystallography*, Holt, Rinehart, and Winston, New York, N. Y. 10001 (1961).
15. L. Kofler and A. Kofler, *Thermomikromethoden*, Wagner, Innsbruck, Austria (1954).
16. Kofler micro hot stage, *Directions for Use*, Technological Service, Arthur H. Thomas Co., Philadelphia, Pa. 19105 (1958 to date). A very good comprehensive treatment.
17. C. D. Felton, Dark-field microscopy, *Analytical Chemistry* **34**, 880 (1962).
18. E. Leitz, Inc., Rockleigh, N. J. 07647.
19. W. C. McCrone, *Applications of Thermal Microscopy*, Mettler Instrument Corp., Box 100, Princeton, N. J. 08540 (22 pp.).
20. Y. Julian and W. C. McCrone, Accurate use of hot stages, *The Microscope* **19**, 225–241 (1971).
21. E. M. Barrall and M. A. Sweeney, Depolarized light intensity and optical microscopy of some mesophase-forming materials, *Molecular Crystals* **5**, 257–271 (1969).
22. F. T. Jones, Fusion techniques in chemical microscopy, *The Microscope* **16**, 37–43 (1968).
23. N. H. Hartshorne, A hot-wire stage and its application, *The Microscope* **23**, 177–190 (1975).
24. Stanton Redcroft, Copper Mill Lane, London SW170BN, England. A new hot stage.
25. E. L. Charsley and D. E. Tolhurst, The application of hot stage microscopy to the study of pyrotechnic systems, *The Microscope* **23**, 227–237 (1975).
26. C. W. Mason and T. G. Rochow, A microscope cold stage with temperature control, *Industrial and Engineering Chemistry* **6**, 367–369 (1934).
27. T. G. Rochow and C. W. Mason, Breaking emulsions by freezing, *Industrial and Engineering Chemistry* **28**, 1296–1300 (1936).
28. Miniature pumps like "Masterflex" connected to a motor such as "Servodyne," Cole–Parmer Instrument Co., 7425 N. Oak Park Ave., Chicago, Ill. 60648 (1977).
29. Greiner Scientific Corp., Bulletin 3-17-164, New York, N. Y. 10013. A portable cold finger.
30. Thermoelectrics Unlimited, Inc., Wilmington, Del. 19809.
31. A. H. Thomas Co., Philadelphia, Pa. 19105.
32. Beckman Instruments, Inc., Cedar Grove, N. J. 07009.
33. E. M. Chamot and C. W. Mason, *Handbook of Chemical Microscopy, Vol. 2: Chemical Methods*, 2nd ed., John Wiley and Sons, Inc., New York, N. Y. 10016 (1940).
34. J. A. Davidson, Pressure cells in optical microscopy, *The Microscope* **23**, 61–71 (1975).
35. H. Reumuth and T. Loske, Kuvette-mikroskopie in biologie und tecknik, *Mikroskopie* **17**, 149–178 (1962), Georg Fromme and Co., Munich 9, Germany.

References for Chapter 12

1. *Compilation of ASTM Standard Definitions*, 3rd ed., American Society for Testing and Materials, Philadelphia, Pa. 19103 (1976).
2. C. E. Hall, *Introduction to Electron Microscopy*, 2nd ed., McGraw-Hill Book Co., Hightstown, N. J., 08520 (1966).
3. E. M. Chamot and C. W. Mason, *Handbook of Chemical Microscopy*, 3rd ed., Vol. 1, John Wiley and Sons, Inc., New York, N. Y. 10016 (1958).
4. G. Rempfer, R. Connell, L. Mercer, and I. Louiselle, An electrostatic transmission microscope, *American Laboratory*, pp. 39–44, 46. (April 1972).
5. B. M. Siegel, *Modern Developments in Electron Microscopy*, Academic Press, New York, N. Y., 10003 (1964).
6. N. D. Yeomans, An Ultrastructural Sonnet, *Journal of Microscopy* **103**, Part 1, p. 131 (January 1975).

7. E. Ruska, Past and present attempts to attain the resolution limit of the transmission microscope, in *Advances in Optical and Electron Microscopy*, Vol. 1 (R. Barer and V. E. Cosslett, eds.), pp. 116–179, Academic Press, New York, N. Y. 10003 (1966).

8. P. R. Swann, C. J. Humphreys, and M. J. Goringe (eds.), High voltage electron microscopy, in *Proceedings of the Third International Conference*, Academic Press, New York, N. Y. 10003 (1973).

9. P. Tucker and R. Murray, A study of amine-etched poly(ethylene terphthalate) filaments by HVEM and SEM, in *33rd Annual Proceedings of the Electron Microscopy Society of America* (G. W. Bailey, ed.), pp. 82–83, Claiter's Publishing Div., Baton Rouge, La. 70821 (1975).

10. A. Septier, The struggle to overcome spherical aberration in electron optics, in *Advances in Optical and Electron Microscopy*, Vol. 1 (R. Barer and V. E. Cosslett, eds.), pp. 204–274, Academic Press, New York, N. Y. 10003 (1966).

11. V. F. Holland, private communication to T. G. Rochow, Monsanto Triangle Park Development Center, Inc., Research Triangle Park, N. C. 27709.

12. A. L. Robinson, Electron microscopy: Imaging molecules in three dimensions, *Science* **192,** 360–362, 400 (1976).

13. E. M. Slayter, *Optical Methods in Biology*, Interscience Div., John Wiley and Sons, Inc., New York, N. Y. 10016 (1970).

14. J. M. Cowley and S. Iijima, Electron microscopy of atoms in crystals, *Physics Today*, pp. 32–40 (March 1977).

15. *Annual Book of ASTM Standards and Index*, American Society for Testing and Materials, Philadelphia, Pa. 19103.

16. M. L. Ladd, *The Electron Microscope Handbook, Specimen Preparation and Related Laboratory Procedures*, Ladd Research Industries, Inc., Burlington, Vt. 05401 (1973).

17. M. A. Hayat, *Basic Electron Microscopy Techniques*, Van Nostrand Reinhold Co., New York, N. Y. 10001 (1972).

18. M. A. Hayat (ed.), *Principles and Techniques of Electron Microscopy Techniques*, 5 vols., Van Nostrand Reinhold Co., New York, N. Y. (1970–1975).

19. R. F. Bils, *Electron Microscopy, Laboratory Manual and Handbook*, Western Publishing Co., Los Angeles, Ca. 90028 (1974).

20. A. M. Glauert (ed.), *Practical Methods in Electron Microscopy*, The Netherlands and American Elsevier Publishing Co., New York, N. Y. 10017, 5 vols. (1972–1977).

21. R. Barer and V. E. Cosslett (eds.), *Advances in Optical and Electron Microscopy*, Academic Press, New York, N. Y. 10003, 6 vols. (to 1976).

22. *ASTM Manual on Electron Metallography Techniques*, Special Technical Publication No. 547, American Society for Testing and Materials, Philadelphia, Pa. 19103 (1973).

23. W. R. Goynes and J. A. Harris, Structural characterization of grafted cotton fibers by microscopy, *Journal of Polymer Science*, Part C, No. **37,** pp. 277–289 (1972).

24. W. C. McCrone, J. G. Delly, and McCrone Associates, *The Particle Atlas*, 2nd ed., Vol. 3, *Electron Microscopy Atlas*, Ann Arbor Science Publishers, Inc., Ann Arbor, Mich. 48106 (1973).

25. R. P. Loveland, *Photomicrography*, Vols. 1 and 2, John Wiley and Sons, Inc., New York, N. Y. 10016 (1970).

26. Some things every electron microscopist ought to know, *Tech Bits No. 65-2*, pp. 3–10, Special Applications Sales, Eastman Kodak Co., Rochester, N. Y. 14650 (1965).

27. Photographic techniques in electron microscopy, *Kodak Pamphlet No. P-109*, pp. 10–70, Scientific Photography Markets, Eastman Kodak Co., Rochester, N. Y. 14650 (1970).

28. T. P. Turnbull, E. F. Fullam, and C. Schaffer, An image intensifier and a Polaroid® camera for the TEM, in *34th Annual Proceedings, Electron Microscopy Society of America*

(G. W. Bailey, ed.), Miami Beach, Fla. (1976). See also sales literature on Catalog No. 8100 Polaroid® Camera System for TEM, Ernest F. Fullam, Inc., P. O. Box 444, Schenectady, N. Y. 12301.

29. R. E. Stevens, Microscopy with electron beams, *American Laboratory*, pp. 49–58 (April 1972).

30. I. W. Drummond, Scanning transmission electron microscopy, *American Laboratory*, pp. 83, 84, 86–90, 92, 94, 95 (April 1976).

31. W. C. McCrone and R. I. Johnson, *Techniques, Instruments and Accessories for Microanalysts*, Walter C. McCrone Associates, Inc., Chicago, Ill. 60616.

References for Chapter 13

1. *Compilation of ASTM Standard Definitions*, 3rd ed., American Society for Testing and Materials, Philadelphia, Pa. 19103 (1976).

2. J. I. Goldstein and H. Yakowitz (eds.), *Practical Scanning Electron Microscopy*, Plenum Press, New York, N. Y. 10011 (1975).

3. W. R. Goynes and J. A. Harris, Structural characterization of grafted cotton fibers by microscopy, *J. Polymer Science*, Part C, No. 37, pp. 277–289 (1972).

4. C. E. Hall, *Introduction to Electron Microscopy*, 2nd ed., McGraw-Hill Book Co., Hightstown, N. J., 08520 (1966).

5. C. W. Oatley, *The Scanning Electron Microscope*, Cambridge University Press, London NW1 2DB (New York, N. Y. 10022) (1972).

6. T. Mulvey and R. K. Webster, eds., *Modern Physical Techniques in Materials Technology*, Oxford University Press, London, W. 1., England (1974).

7. J. W. S. Hearle, J. T. Sparrow, and P. M. Cross, *The Use of the Scanning Electron Microscope*, Pergamon Press, Elmsford, N. Y. 10523 (1972).

8. T. E. Everhart and T. L. Hayes, The scanning electron microscope, *Scientific American* **226**, 55–69 (January 1972).

9. O. C. Wells, *Scanning Electron Microscopy*, McGraw–Hill Book Co., New York, N. Y. 10020 (1974).

10. A. C. Reimschuessel, Scanning electron microscopy, *J. Chemical Education* **49**, A413–A449 (1972).

11. E. M. Chamot and C. W. Mason, *Handbook of Chemical Microscopy*, Vol. 1, 3rd ed., John Wiley and Sons, Inc., New York, N. Y. 10016 (1958).

References for Chapter 14

1. E. W. Müller, Field ion microscopy, *Science* **149**, No. 3684, 591–601 (August 6, 1965).

2. C. E. Hall, *Introduction to Electron Microscopy*, 2nd ed., McGraw-Hill Book Co., Hightstown, N. J., 08520 (1966).

3. B. Ralph, Field-ion microscopy, Chapter 5, in *Modern Physical Techniques in Materials Technology* (T. Mulvey and R. K. Webster, eds.), Oxford University Press, London, W.1 (1974).

4. R. Gomer, *Field Emission and Field Ionization*, Harvard University Press, Cambridge, Mass. 02138 (1961).

5. *Compilation of ASTM Standard Definitions*, 3rd ed., American Society for Testing and Materials, Philadelphia, Pa. 19103 (1976).

6. Field-ion microscope, Materials Research Corp., Orangeburg, N. Y. 10962. Private communication, December 15, 1965.

7. S. J. Fonash, Advances in the understanding of image formation in FIM, *Microstructures*, pp. 17–21 (August/September 1972).
8. J. J. Hren and S. Ranganathan (eds.), *Field-Ion Microscopy*, Plenum Publishing Corp., New York, N. Y. 10060 (1968).
9. E. W. Müller and T. T. Tsong, Field-Ion Microscopy, Elsevier, New York, N. Y. 10017 (1969).
10. R. F. Hockman, E. W. Müller, and B. Ralph, *Applications of Field-Ion Microscopy*, Georgia Technical Press, Atlanta, Ga. 30332 (1969).
11. ASTM National Committee E-4 on Metallography, organizational information in the annual *Yearbook*, American Society for Testing and Materials, Philadelphia, Pa. 19103.

References for Chapter 15

1. *Compilation of ASTM Standard Definitions*, 3rd ed., American Society for Testing and Materials, Philadelphia, Pa. 19103 (1976).
2. CANALCO leaflet, The Micro-Source of X-Rays, Canal Industrial Corp., Bethesda, Md. 20014 (1959).
3. S. B. Newman and D. Fletcher, Soft-x-ray microscopy of paper, *Technical Association of the Pulp and Paper Industry* **47**, 177–180 (1964).
4. M. C. Botty and E. J. Thomas, Applications of microradiography with the EMU-1 electron microscope, paper presented to the Electron Microscope Society of America, Columbus, Ohio, September 9, 1959 (unpublished). M. C. Botty and F. G. Rowe, U. S. Patent 2,843,751 (July 15, 1958).
5. Philips Electronic Instruments, Mount Vernon, N. Y. 10550.
6. M. C. Botty, unpublished work, American Cyanamid Co., Stamford, Conn. 06904; private communications of May 21 and June 2, 1976 to T. G. Rochow.
7. J. I. Gedney, M. C. Botty, and E. J. Thomas, The preparation of rigid foams for microscopical examination, paper presented to the American Society for Testing and Materials, Atlantic City, N. J., June 29, 1966 (unpublished).
8. W. C. Nixon, Chapter on x-ray microscopy, in *Modern Methods of Microscopy* (A. E. J. Vickers, ed.), Butterworth Scientific Publications, London, England (1956).
9. E. P. Bertin and R. J. Longobucco, Practical x-ray contact microradiography, Part 1, *RCA Scientific Instrument News* **5**, 4 (1960); and Part 2, *RCA Scientific Instrument News* **6**, 1 (1961), Radio Corporation of America, Camden, N. J. 08100.
10. E. M. Slayter, *Optical Methods in Biology*, Interscience Div., John Wiley and Sons, Inc., New York, N. Y. 10016 (1970).

References for Chapter 16

1. C. F. Quate, Acoustic microscopy, *Trends in Biochemical Sciences;* see section on emerging techniques, N127-N129 (June 1977).
2. D. White (ed.), Vol. 1; D. White and R. Barnes (eds.), Vol. 2; D. White and R. Brown (eds.), Vols. 3A and 3B, *Ultrasound in Medicine*, Plenum Publishing Corp., New York, N. Y. 10011 (1975, 1976, and 1977).
3. Sales literature of Sonoscan®, Inc., 720 Foster Ave., Bensenville, Ill. 60106 (1977).
4. R. A. Lemons and C. F. Quate, Acoustic microscopy: Biomedical applications, *Science* **188**, 905–911 (1975).
5. R. Kompfner and C. F. Quate, Acoustic radiation as used for microscopy, *Physics in Technology* (November 1977).
6. C. F. Quate, Scanning Acoustic Microscope, material prepared for the Symposium

Workshop, R. C. Eggleton, Director Ultrasound Research Labs, Indianapolis Center for Advanced Research, Inc., Indianapolis, Ind. 46202 (February 4–18, 1977).

7. C. F. Quate, Edward L. Ginzton Laboratory, Stanford University, Stanford, Ca. 94305.

8. Information distributed at the First Acoustic Microscopy Symposium Workshop at the Indianapolis Center for Advanced Research, Indianapolis, Ind. 46202, February 14–18, 1977.

9. L. W. Kessler, High resolution visualization of tissue with acoustic microscopy, in *Proceedings of the Second World Congress on Ultrasonics in Medicine*, Rotterdam, June 1973, M. deVlieger (ed.), Excerpta Medica Foundation (1974).

10. L. W. Kessler, Acoustic microscopy: Progress and applications 1977, ASA 94th Meeting, Miami Beach, Fla., December 1977.

11. L. W. Kessler, A. Korpel, and P. R. Palermo, Simultaneous acoustic and optical microscopy of biological specimens, *Nature* **239**, 111–112 (September 8, 1972).

12. A. Madeyski and L. W. Kessler, Initial experiments in the application of acoustic microscopy to the characterization of steel and to the study of fracture phenomena, *IEEE Transactions on Sonics and Ultrasonics*, **SU-23**, No 5 (September 1976).

13. L. W. Kessler and D. E. Yuhas, Acoustical microscopy—A new tool for materials analysis and NDT, 37th National ASNT Fall Conference Proceedings October 3–6, 1977.

14. R. C. Eggleton and F. S. Vinson, Heart model supported in organ culture and analyzed by acoustic microscopy, *Acoustical Holography Vol.7—Recent Advances in Ultrasonic Visualization* (L. W. Kessler, ed.), Plenum Press, New York, N. Y. 10011 (1977).

Author Index

Subject Index